Kawasaki 250, 350 and 400 3 Cylinder Models Owners Workshop Manual

by Frank Meek

Models covered:

S1 Series. 249 cc. UK 1972-1976 USA 1971-1975
S2 Series. 346 cc. UK and USA 1972 - 1973
S3 Series. 400 cc. UK 1974-1976 USA 1974-1975
KH 250. 249 cc. UK 1976 - 1979 USA 1975- 1979
KH 400. 400 cc. UK 1976 - 1979 USA 1975- 1979

ISBN 978 0 85696 134 2

© Haynes Group Limited 1989

(134-805)

Haynes Group Limited
Haynes North America, Inc

www.haynes.com

Acknowledgements

Grateful thanks are due to Kawasaki Motors (UK) Limited for permission to use certain of their drawings and for the technical assistance given so freely when this manual was being prepared.

Brian Horsfall dismantled and reassembled the model used for the photographs in this manual and devised the ingenious methods for overcoming the lack of service tools.

Les Brazier arranged and took the photographs. Jeff Clew edited the text.

The cover photograph was arranged through the courtesy of Kawasaki (UK) Ltd.

The machine used for the production of this manual was supplied by Bryan Goss Motor Cycles of Yeovil.

We would also like to thank the Avon Rubber Company who kindly supplied the illustrations and advice about tyre fitting.

Finally we thank Ken Heanes Ltd and especially Dave Lewendon, the manager of the Aldershot Branch, for much assistance with Kawasaki spare parts and technical advice.

About this manual

This manual contains a detailed stripdown and reassembly procedure for the major assemblies of the 'S' series Kawasaki three cylinder two-strokes.

The machine owner is advised that where only part of a stripdown is required the manual should be studied to see what parts of the machine require removal. This will save time, labour and probably expense.

With the help of this manual the owner should find maintaining his machine an interesting hobby and in these days of soaring garage charges could save himself the money and trouble of taking his machine to be serviced by others.

In the main, the manufacturer's service tools have not been called for. Alternative methods of loosening or slackening some vital component are given whenever possible. If it is essential to obtain a tool, your dealer may have a system for lending the item against a deposit.

Each of the six chapters is divided into numbered sections. Within each section are numbered paragraphs. Cross reference throughout this manual is quite straightforward and logical. When reference is made "See Section 6.10" it simply means you should read in the same chapter, Section 6, paragraph 10. If reference is made to another chapter say Chapter 2, then the reference would read "See Chapter 2, Section 6.10".

All photographs are captioned with a section/paragraph number to which they refer and are always relevant to the chapter text adjacent.

Figure numbers appear in numerical order within a chapter. Figure 1.1 is therefore the first figure in Chapter 1.

Left hand and right hand descriptions of a machine and a component refer to the left and right of a machine with the rider seated normally.

Motorcycle manufacturers continually make changes to specifications and recommendations, and these, when notified, are incorporated into our manuals at the earliest opportunity.

Whilst every care is taken to ensure that the information in this manual is correct, no liability can be accepted by the authors or publishers for loss, damage or injury caused by any errors in or omissions from the information given.

Contents

1976 250 cc KH 250 model

1976 400 cc KH 400 model

Introduction to the Kawasaki 'S' series machines

The 'S' series machines dealt with in this manual are the 250 cc S1, the 350 cc S2 and the 400 cc S3. They are all two-stroke, triple cylinder models. The range of models covered is as follows:

S1, S1.A, S1.B (1971 onwards)
S2, S2.A (1973 - 1974)
S3 (1974 onwards)

The models are part of the range of motorcycles manufactured by Kawasaki Heavy Industries Ltd (motorcycle division) and are marketed in the USA, UK and Europe.

Following the normal procedure at Sparkford it was decided to purchase, strip down completely and rebuild a used machine to give an idea of how a machine will look to the motorcyclist carrying out his own servicing and repairs. All the photographs are of a 1972 S2 model.

Where possible modifications and variations in the different models are described in the text.

Kawasaki Dimensions

(European Models)

Dimension	S1 Series	S2 Series	S3 Series
Overall length	80.9 inch (2.055 mm)	78.7 inch (2.000 mm)	80.9 inch (2.055 mm)
Overall width	29.9 inch (760 mm)	30.3 inch (770 mm)	29.9 inch (760 mm)
Overall height	40.7 inch (1.035 mm)	41 inch (1.045 mm)	41.1 inch (1.045 mm)
Wheelbase	54.1 inch (1.375 mm)	51.5 inch (1.310 mm)	53.7 inch (1.365 mm)
Ground clearance	5.9 inch (150 mm)	6.3 inch (160 mm)	5.9 inch (150 mm)
Dry weight	339 lbs (155 kg)	335 lbs (152 kg)	353 lbs (159 kg)

Ordering spare parts

When ordering spare parts for your machine it is advisable to deal directly with an accredited Kawasaki dealer who should be able to supply you with parts from stock.

Always quote the engine and frame numbers in full, particularly if you have an older machine. The frame number is stamped on the steering head of the frame. The engine number is stamped on the crankcase adjacent to the crankcase oil dipstick. Quote the full model number of the machine and, where necessary, the colour scheme.

Use only parts of genuine Kawasaki supply. Pattern parts are available, some of which originate in the country of the machine's manufacture and are packed in an almost identical container to the genuine article. These replacement parts do not necessarily make a satisfactory replacement although they are cheaper initially.

Some of the more expendable parts such as spark plugs, bulbs, tyres, oils and greases etc, can be obtained from accessory shops and motor factors, who have convenient opening hours, charge lower prices and can often be found not far from home. It is also possible to obtain parts on a Mail Order basis from a number of specialists who advertise regularly in the motor cycle magazines.

Engine No. Location

Frame No. Location

Routine maintenance

Periodic routine maintenance is a continuous process that commences immediately the machine is used and continues until the machine is no longer fit for service. It must be carried out at specified mileage recordings or on a calendar basis if the machine is not used regularly, whichever is the soonest. Maintenance should be regarded as an insurance policy, to help keep the machine in the peak of condition and to ensure long, trouble-free service. It has the additional benefit of giving early warning of any faults that may develop and will act as a safety check, to the obvious advantage of both rider and machine alike.

The various maintenance tasks are described under their respective mileage and calendar headings. Accompanying photos or diagrams are provided, where necessary. It should be remembered that the interval between the various maintenance tasks serves only as a guide. As the machine gets older, is driven hard, or is used under particularly adverse conditions, it is advisable to reduce the period between each check.

If a specific item is mentioned but not described in detail it will be covered fully in the appropriate chapter. No special tools are required for the normal routine maintenance tasks. The tools contained in the tool kit supplied with every new machine will suffice but if they are not available, the tools found in the average garage will make an adequate substitute.

Weekly or every 200 miles

1 Check tyre pressures when tyres are cold.
2 Check the level of the electrolyte in battery.
3 Check the oil in oil tank for the engine oil and with the crankcase dipstick for the gearbox oil.
4 Give the machine a close visual inspection, checking for loose nuts and fittings, frayed control cables, etc.
5 Ensure the lights, horn and traffic indicators function correctly, also the speedometer.

One month or every 500 miles

1 Check and tighten all nuts and bolts.
2 Carefully examine the tyres for faults, damage, etc.
3 Check and tighten the spokes if required.
4 Re-adjust the brakes if required.
5 Re-adjust the clutch release if required.

6 Clean, adjust and oil the chain.
7 Replace the transmission oil (after first 500 miles then at 2000 mile checks).
8 Clean and reset the spark plug gaps.
9 Clean the points and check the ignition timing.
10 Check the oil pump and re-align if necessary.
11 Check the carburettors and reset the controls if necessary.
12 Check the battery voltage and top up with distilled water, if required.

Three months or every 2000 miles

1 Rectify any faults that are outstanding.
2 Check the front fork oil.
3 Clean the fuel tap filter.
4 Clean the petrol and engine oil tanks.
5 Clean and adjust the spark plugs.
6 Clean the points and check the ignition timing.
7 Clean the air cleaner element.
8 Check the battery voltage and top up with distilled water if necessary.
9 Drain and change the gearbox oil.
10 Examine the clutch release and reset if required.
11 Check the steering for play.
12 Check wheel alignment.
13 Examine the chain for wear. Clean and oil it.
14 Adjust the brakes if necessary.
15 Examine and tighten the spokes if necessary.
16 Carefully examine the tyres for damage.
17 Check and tighten the nuts and bolts on the machine.
18 Check the oil pump and re-adjust if necessary.
19 Adjust the carburettor controls if necessary.

Six months or every 4000 miles

1 Rectify any faults outstanding.
2 Replace the spark plugs.
3 Lubricate at all necessary points. (Refer to Lubrication Chart)
4 Decarbonise the exhaust system.
5 Decarbonise the heads and ports.
6 Examine and clean the brake shoes.
7 Check the wear on both sprockets.

Summary
of routine maintenance adjustments and capacities

Item	S1	S2	S3
Spark plug gaps: inches	0.016 - 0.020	As S1	As S1
mm	0.4 - 0.5		
Spark plug types: Standard	NGK B9HCS	As S1	As S1
Alternative	Champion L77J		
Contact breaker gaps: inches	0.012 - 0.016	As S1	As S1
mm	0.3 - 0.4		
Ignition timing: Degrees BTDC	23	As S1	As S1
Tyre pressures: Front	24 psi	As S1	As S1
Rear	31 psi		
Add 4 psi to rear if luggage or passenger carried			
Fuel tank capacity: Imp galls	3.08	As S1	As S1
Litres	14		
US galls	3.7		
Oil tank capacity (Engine) Imp qts	1.32	As S1	As S1
2 stroke oil Litres	1.5		
US qts	1.6		
Gearbox oil capacity: (SAE 10W30 or Imp qts	0.97	As S1	As S1
10W40) Litres	1.1		
US qts	1.16		
		1973 on	1973 on
Front fork oil ozs	7.1	10.48	As S2
(SAE 10W non-detergent) cc	210	310	
Oil level from top of tube: ins	14.75	13.98	
mm	375	355	

Torque settings

Part	Quantity	ft lb	kg m
Front wheel spindle	1	48 - 61	6.7 - 8.5
Transmission oil drain plug	1	36.9 - 52.1	5.1 - 7.2
Engine mounting bolts	4	3 - 4	0.415 - 0.533
Spokes	40	1.836 - 2.166	0.25 - 0.3
Spark plugs	3	18.1 - 21.7	2.5 - 3.0
Cylinder head nuts	4	16	2.2
Crankcase (6 mm)	12	10.8 - 11.6	1.5 - 1.6
Crankcase (8 mm)	8	15.9 - 19.5	2.2 - 2.7
Rear wheel spindle	1	48 - 61	6.7 - 8.5
Swinging arm pivot shaft	1	78.1 - 108	10.8 - 15
Disc bolts	2	11.6 - 16	1.6 - 2.2
Spindle clamp bolts	4	13 - 14.5	1.8 - 2.0

Recommended lubricants and fluids

COMPONENT	TYPE OF LUBRICANT	RECOMMENDED GRADE
Gearbox oil	SAE 10W30 or 10W40	Castrolite
Front forks	SAE 10W - non detergent	Castrol Hyspin
Hydraulic disc brakes	Fluid to DOT 3	Castrol Girling Universal Brake and Clutch Fluid
Chain	SAE 30	Linklyfe or Chainguard
Engine oil	Two-stroke engine oil	Castrol TT Two-Stroke Oil
All lubrication points, control cables etc	Light oil 10W/30 Multigrade	Castrolite
All grease points, cam spindles etc	Lithium based high melting point grease	Castrol LM Grease

Unit conversion table

Unit	Multiply by	Equals
cc	0.0610	cu in
cc	0.02816	fluid oz (Imp)
cc	0.03381	fluid oz (US)
cu ins	16.39	cc
fluid oz (Imp)	35.51	cc
fluid oz (US)	29.57	cc
ft lbs	0.1383	kg m
gallons (Imp)	4.546	litres
gallons (Imp)	1.201	gallons (US)
gallons (US)	3.7853	litres
gallons (US)	0.8326	gallons (Imp)
grams	0.03527	oz
inches	25.40	mm
inch pounds	0.0833	ft lbs
inch pounds	0.0115	kg metres
kg	2.2046	lbs
kg	35.274	oz
kg metre	7.233	ft lbs
kg metre	86.796	inch pounds
kg cm^2	14.22	lbs sq inch
kg	0.6214	miles
lb	0.4536	kg
lb/sq in	0.0703	kg/cm^2
litre	28.16	fluid oz (Imp)
litre	33.81	fluid oz (US)
litre	0.8799	quart (Imp)
litre	1.0567	quart (US)
metre	3.281	ft
mile	1.6093	km
mm	0.03937	in
oz	28.35	grams
qt (Imp)	1.1365	litres
qt (Imp)	1.201	qt (US)
qt (US)	0.9463	litres
qt (US)	0.8326	qt (Imp)

Common abbreviations

ABDC	After bottom dead centre
ATDC	After top dead centre
BBDC	Before bottom dead centre
BDC	Bottom dead centre
BTDC	Before top dead centre
cc	Cubic centimetres
cu in	Cubic inches
fl oz	Fluid ounces
ft	Feet
ft lbs	Foot pounds
gal	Gallons
hp	Horse power
in lbs	Inch pounds
kg	Kilogram
kg cm^2	Kilograms per square centimetre
kg m	Kilogram metre
km	Kilometre
kph	Kilometre per hour
l	Litre
mm	Millimetres
psi	Pounds per square inch
qt	Quart
rpm	Revolutions per minute
SS	Standing start
"	Inches
Imp	Imperial standard
US	United States standard

Safety first!

Professional motor mechanics are trained in safe working procedures. However enthusiastic you may be about getting on with the job in hand, do take the time to ensure that your safety is not put at risk. A moment's lack of attention can result in an accident, as can failure to observe certain elementary precautions.

There will always be new ways of having accidents, and the following points do not pretend to be a comprehensive list of all dangers; they are intended rather to make you aware of the risks and to encourage a safety-conscious approach to all work you carry out on your vehicle.

Essential DOs and DON'Ts

DON'T start the engine without first ascertaining that the transmission is in neutral.

DON'T suddenly remove the filler cap from a hot cooling system – cover it with a cloth and release the pressure gradually first, or you may get scalded by escaping coolant.

DON'T attempt to drain oil until you are sure it has cooled sufficiently to avoid scalding you.

DON'T grasp any part of the engine, exhaust or silencer without first ascertaining that it is sufficiently cool to avoid burning you.

DON'T allow brake fluid or antifreeze to contact the machine's paintwork or plastic components.

DON'T syphon toxic liquids such as fuel, brake fluid or antifreeze by mouth, or allow them to remain on your skin.

DON'T inhale dust – it may be injurious to health (see *Asbestos* heading).

DON'T allow any spilt oil or grease to remain on the floor – wipe it up straight away, before someone slips on it.

DON'T use ill-fitting spanners or other tools which may slip and cause injury.

DON'T attempt to lift a heavy component which may be beyond your capability – get assistance.

DON'T rush to finish a job, or take unverified short cuts.

DON'T allow children or animals in or around an unattended vehicle.

DON'T inflate a tyre to a pressure above the recommended maximum. Apart from overstressing the carcase and wheel rim, in extreme cases the tyre may blow off forcibly.

DO ensure that the machine is supported securely at all times. This is especially important when the machine is blocked up to aid wheel or fork removal.

DO take care when attempting to slacken a stubborn nut or bolt. It is generally better to pull on a spanner, rather than push, so that if slippage occurs you fall away from the machine rather than on to it.

DO wear eye protection when using power tools such as drill, sander, bench grinder etc.

DO use a barrier cream on your hands prior to undertaking dirty jobs – it will protect your skin from infection as well as making the dirt easier to remove afterwards; but make sure your hands aren't left slippery. Note that long-term contact with used engine oil can be a health hazard.

DO keep loose clothing (cuffs, tie etc) and long hair well out of the way of moving mechanical parts.

DO remove rings, wristwatch etc, before working on the vehicle – especially the electrical system.

DO keep your work area tidy – it is only too easy to fall over articles left lying around.

DO exercise caution when compressing springs for removal or installation. Ensure that the tension is applied and released in a controlled manner, using suitable tools which preclude the possibility of the spring escaping violently.

DO ensure that any lifting tackle used has a safe working load rating adequate for the job.

DO get someone to check periodically that all is well, when working alone on the vehicle.

DO carry out work in a logical sequence and check that everything is correctly assembled and tightened afterwards.

DO remember that your vehicle's safety affects that of yourself and others. If in doubt on any point, get specialist advice.

IF, in spite of following these precautions, you are unfortunate enough to injure yourself, seek medical attention as soon as possible.

Asbestos

Certain friction, insulating, sealing, and other products – such as brake linings, clutch linings, gaskets, etc – contain asbestos. *Extreme care must be taken to avoid inhalation of dust from such products since it is hazardous to health.* If in doubt, assume that they *do* contain asbestos.

Fire

Remember at all times that petrol (gasoline) is highly flammable. Never smoke, or have any kind of naked flame around, when working on the vehicle. But the risk does not end there – a spark caused by an electrical short-circuit, by two metal surfaces contacting each other, by careless use of tools, or even by static electricity built up in your body under certain conditions, can ignite petrol vapour, which in a confined space is highly explosive.

Always disconnect the battery earth (ground) terminal before working on any part of the fuel or electrical system, and never risk spilling fuel on to a hot engine or exhaust.

It is recommended that a fire extinguisher of a type suitable for fuel and electrical fires is kept handy in the garage or workplace at all times. Never try to extinguish a fuel or electrical fire with water.

Note: *Any reference to a 'torch' appearing in this manual should always be taken to mean a hand-held battery-operated electric lamp or flashlight. It does* **not** *mean a welding/gas torch or blowlamp.*

Fumes

Certain fumes are highly toxic and can quickly cause unconsciousness and even death if inhaled to any extent. Petrol (gasoline) vapour comes into this category, as do the vapours from certain solvents such as trichloroethylene. Any draining or pouring of such volatile fluids should be done in a well ventilated area.

When using cleaning fluids and solvents, read the instructions carefully. Never use materials from unmarked containers – they may give off poisonous vapours.

Never run the engine of a motor vehicle in an enclosed space such as a garage. Exhaust fumes contain carbon monoxide which is extremely poisonous; if you need to run the engine, always do so in the open air or at least have the rear of the vehicle outside the workplace.

The battery

Never cause a spark, or allow a naked light, near the vehicle's battery. It will normally be giving off a certain amount of hydrogen gas, which is highly explosive.

Always disconnect the battery earth (ground) terminal before working on the fuel or electrical systems.

If possible, loosen the filler plugs or cover when charging the battery from an external source. Do not charge at an excessive rate or the battery may burst.

Take care when topping up and when carrying the battery. The acid electrolyte, even when diluted, is very corrosive and should not be allowed to contact the eyes or skin.

If you ever need to prepare electrolyte yourself, always add the acid slowly to the water, and never the other way round. Protect against splashes by wearing rubber gloves and goggles.

Mains electricity

When using an electric power tool, inspection light etc which works from the mains, always ensure that the appliance is correctly connected to its plug and that, where necessary, it is properly earthed (grounded). Do not use such appliances in damp conditions and, again, beware of creating a spark or applying excessive heat in the vicinity of fuel or fuel vapour.

Ignition HT voltage

A severe electric shock can result from touching certain parts of the ignition system, such as the HT leads, when the engine is running or being cranked, particularly if components are damp or the insulation is defective. Where an electronic ignition system is fitted, the HT voltage is much higher and could prove fatal.

Chapter 1 Engine, clutch and gearbox

Contents

Specifications

	S1B	S2A	S3
Engine			
Type	2 stroke, 3 cylinder piston	As S1B	As S1B
Bore/stroke	1.77 x 2.06 in	2.09 x 2.06 in	2.24 x 2.06 in
Displacement	249 cc	346.2 cc	400.4 cc
Compression ratio	7.5 : 1	7.3 : 1	6.5 : 1
Max hp	28 ps @ 7500 rpm	44 ps @ 8000 rpm	42 ps @ 7000 rpm
Max torque	19.5 ft lb @ 7000 rpm	28.9 ft lb @ 7500 rpm	31.2 ft lb @ 6500 rpm
Port timing			
Intake - open	74º BTDC	73º BTDC	As S2A
close	74º ATDC	73º ATDC	As S2A
Scavenge - open	62º BBDC	58º BBDC	As S2A
- close	62º ABDC	58º ABDC	As S2A
Exhaust - open	83º BBDC	89º BBDC	86º BBDC
- close	83º ABDC	89º ABDC	86º ABDC
Engine oil	2 stroke oil	2 stroke oil	2 stroke oil

Transmission

Type ...	...	...	...	...	...	...	...	5 speed constant mesh return speed	As S1B	As S1B

Clutch ... ...   Heavy duty, wet plate, multiple disc   As S1B   As S1B

Friction plate thickness:
Standard ...	...	...	...	...	...	...	...	0.118 in (3 mm)	As S1B	As S1B
Replace at	...	...	...	...	...	...	...	0.106 in (2.7 mm)	As S1B	As S1B

Housing/plate clearance
Standard ... ...   0.0020 - 0.0177 in (0.05 - 0.45 mm)   As S1B   As S1B

Gear ratio
1st	...	...	...	...	...	...	...	...	2.86 (40/14)	As S1B	As S1B
2nd	...	...	...	...	...	...	...	...	1.79 (34/19)	As S1B	As S1B
3rd	...	...	...	...	...	...	...	...	1.35 (31/23)	As S1B	As S1B
4th	...	...	...	...	...	...	...	...	1.12 (28/25)	As S1B	As S1B
5th	...	...	...	...	...	...	...	...	0.96 (26/27)	As S1B	As S1B
Primary reduction ratio	...	...	...	...	...	...		2.22 (60/27)	As S1B	As S1B
Final reduction ratio	...	...	...	...	...	...		3.45 (48/14)	3.07 (43/14)	2.93 (41/14) or 2.64 (37/14)
Overall drive ratio	...	...	...	...	...	...		7.31 (5th gear)	6.56 (5th gear)	6.25 or 5.63

Gearbox oil ... ...   SAE 10W30 or 40   As S1B   As S1B
Gearbox oil capacity ... ...   0.97 Imp qts (1.1 litre)   As S1B   As S1B

Note: specifications liable to alteration without notice

1 General description

The engine/gearbox fitted to the 'S' series Kawasaki models is of the two stroke type employing three cylinders. It uses aluminium alloy pistons and loop scavenging to achieve a better induction and exhaust efficiency.

Wide fins are fitted to the outside of the cylinders to achieve greater cooling power. The S3 machine can easily be identified by the square shaped profile of these fins.

The crankcase separates into lower and upper sections and gives adequate access to the gearbox components. It is impossible to work on the gearbox components or crankshaft without removing and separating the crankcases.

The gearbox has five speeds and is fitted with a conventional kickstarter. Primary drive is through a pair of helically cut pinions via a multiplate clutch. A positive stop mechanism is incorporated in the gearchange mechanism to ensure each gear is selected with positive action.

The 'S' series machines use the Superlube system of oil lubrication. The Superlube uses oil from a separate tank which gravity feeds an engine-driven oil pump which in turn meters oil to the cylinder induction tracts. The amount of oil fed to the engine is controlled by engine speed and also via the throttle control, so that oil delivery relates to changing operating conditions.

The crankcase has its own separate oil content for gearbox lubrication and is fitted with a dipstick for monitoring the contents. It holds just short of an Imperial quart and uses SAE 10W/30 or 10W/40 oil.

2 Operations with engine/gearbox in frame

1 It is not necessary to remove the engine unit from the frame unless the crankshaft assembly or the gearbox bearings or pinions require attention. Most operations can be accomplished with the engine in place.
2 A list of these items is given herewith:
3 Air cleaner — Remove left side cover then air inlet pipes for access.
4 Carburettors — Remove air inlet pipes for access.
5 Clutch cable and engine sprocket — Remove gearchange pedal, front chain cover, clutch cable, chain for access.
6 AC generator — Remove left engine cover, generator cables, stator, rotor.
7 Right engine cover and oil pump — Drain transmission oil then remove oil pump cover, tachometer cable, oil pump cable, oil inlet pipe, oil outlet pipes, right hand engine cover, oil pump.
8 Clutch and primary gear — Remove pressure plate, friction plates, clutch plates, pressure plate lifter, oil pump pinion, primary drive pinion, clutch centre, clutch housing.
9 Pistons — Remove exhaust pipes, cylinder heads, cylinders, gudgeon pins, pistons, small end bearings, piston rings.

3 Method of engine/gearbox removal

1 It is necessary to remove the engine/gearbox for the following operations:
2 Removal and replacement of the main bearings.
3 Removal and replacement of the gear cluster, selectors and gearbox main bearings.
4 Removal and replacement of the crankshaft assembly.
5 The engine and gearbox are an integral unit and it is necessary to remove the unit complete in order to gain access to either component.
6 Separation is accomplished after the unit has been removed and refitting cannot take place until the crankcases have been assembled.
7 Do not omit to have receptacles for the parts you remove, small or large, heavy or light.
8 Do not neglect all the safety precautions and have strict regard for cleanliness.
9 If on occasions you find a removal procedure is repeated from another section it is to ensure that if you are working in a different order to that recommended you may still be able to proceed.
10 It is emphasised that you should plan the job before you commence, especially if it means taking out the engine. Make sure you have the cleaning gear, tools and replacement parts available. Do not tackle any job beyond your capabilities - it will cost money to rectify.

4 Removing the engine/gearbox unit

1 Place the machine on the centre stand and ensure it is standing firmly.
2 Remove the crankcase drain plug and drain oil from the crankcase. It will assist this stage if the engine has just previously been run and the oil is warm.
3 Disconnect the battery by removing the positive lead. The battery is located beneath the dualseat which will hinge to the left to give access. Undo the securing strap if the battery has to be removed for service. If the battery is left in situ, tape up the loose positive spade terminal.

4 Turn the fuel tap to the S position, to stop the flow of petrol and remove the fuel pipe connection to the carburettors.

5 Lift out the tank, taking full precautions against damaging the main wiring harness located under the fuel tank, between tank and frame.

6 Drain the tank into a clean petrol can by turning the tap to Reserve. Stow the can and tank in a safe location, away from naked lights.

7 Disconnect the nuts securing the exhaust pipe holders to the cylinder studs and lift away each pipe. Slide the holders back over the exhaust pipes. Slacken rear silencer mounting bolts. Remove exhaust system complete (S3 model).

8 Remove the oil pump cover.

9 Uncouple and remove the tachometer cable entering the front of the oil pump compartment.

10 Uncouple the carburettor air inlet hoses by slackening the retaining clamps.

11 Remove the carburettors as complete units. See Chapter 2, Section 5 for details.

12 Remove the oil pump cable at the oil pump. See Chapter 2, Section 14.4.

13 Remove the oil inlet pipe at the oil pump. Squeeze the pipe with the fingers of the left hand to prevent leakage at the point of disconnection. If the end of the pipe is now lifted to saddle height it will not leak oil. The end can be taped and tied, or blocked with a bolt or screw.

14 Remove the gearchange pedal and its associated linkage assembly. Mark the pedal position to make refitting simpler.

15 Remove the crosshead screws securing the chain cover and remove the cover.

16 Remove the chain master link and remove the drive chain. Note carefully the direction of the link.

17 The clutch release assembly is near the engine sprocket at the opposite end of the crankcase from the clutch assembly. Loosen the clutch release locknut and turn the adjusting screw to allow play in the cable.

18 Straighten the clutch release lever tongue and remove the inner section of the clutch cable from the clutch release lever.

19 Ensure the alternator cables are disconnected and examine the engine area carefully to ensure all the electrical joints have been separated.

20 Detach the engine mounting nuts, the mounting bolts or studs, and then the mounting brackets. Take note of any engine mounting shims fitted. Mark and refit them in the original position.

21 Lift the engine out carefully from the right. Remember the engine is a heavy unit so have a bench or engine stand ready to place it on.

22 Extra care must be taken to support the engine when the mounting bolts are removed. Although the Kawasaki engines are not unduly heavy it is advisable to have a second person available during the removal operation to help with the initial engine removal and to steady the frame and cycle parts as the engine is lifted clear.

23 Later versions of the S3 are fitted with eight rubber bushed engine mountings to the frame. This is to reduce engine vibration to the seat and handlebars. These rubber mountings must be carefully inspected for damage or deterioration before refitting to the machine. They do not impede engine removal.

5 Dismantling the engine/gearbox unit - general

1 Before commencing work on the engine unit, the external surfaces should be cleaned thoroughly. A motorcycle has very little protection from road grit and other foreign matter which sooner or later will find its way into the dismantled engine if this simple precaution is not carried out.

2 The demonstration model had in fact been ridden a few hours previously and careful examination of the illustrations will show what to expect during your own servicing.

3 One of the proprietary cleaning compounds such as Gunk or Jizer can be used to good effect, especially if the compound is first allowed to penetrate the film of grease and oil before it is washed away. In the USA Gumout degreaser is an alternative.

3.7 Have receptacles for the parts you remove

4.3 Remove the positive lead

4.4. Remove the fuel connections to the carburettor

4.7 Disconnect the nuts

4.7a Lift away each pipe

4.7b Slide the holders back

4.10 Uncouple the carburettor air inlet tubes

4.20a Detach the engine mounting nuts

4.20b The mounting bolts

4   It is essential when washing down to make sure that water does not enter the carburettors or the electrics particularly now that these parts are more vulnerable.

5   Collect together an adequate set of tools in addition to those of the tool roll carried at the rear of the seat.

6   If the engine has not been previously dismantled, you will need an impact screwdriver. This will safeguard the heads of the crosshead screws used for engine assembly. These are invariably machine-tightening during manufacture. **Caution** - Use great care as the screws and cases are easily damaged. Use a crosshead type screwdriver and NOT one of the Phillips type, which will slip out of the screws.

7   Avoid force in any of the operations. There is generally a good reason why an item is sticking. It is probably due to your using the wrong procedure or sequence of operations.

8   Dismantling will be made easier if a simple engine stand is constructed that will correspond with the engine mounting joints. This arrangement will permit the complete unit to be clamped rigidly to the workbench, leaving both hands free for dismantling.

6 Dismantling the engine/gearbox - removing the cylinder head

1   Remove the four nuts holding each cylinder head in position. Carefully lift each head upwards taking care not to damage the threads of the holding down studs in the process.

2   Spark plugs can be left in situ in the heads during engine overhaul if desired but check them for correct type, condition and gap before replacing the cylinder head.

7 Dismantling the engine/gearbox - removing the cylinder barrels, pistons and rings

1   After removing the cylinder heads and spark plugs tap the cylinder barrels lightly with a wooden block around the exhaust port and lift them off the crankcase.

2   Take care to support each piston as it emerges from the cylinder bore to avoid damage.

3   To avoid entry of any dirt or dust into the crankcase at this stage, it is advisable to stuff the crankcase mouth with clean cloths. This will prevent particles of piston ring from falling in if the rings are broken and there is no need to strip the engine completely.

4   Remove the circlips from each piston boss and discard them. Circlips should never be re-used.

5   Using a drift of the correct diameter (if necessary) tap each gudgeon pin out of the piston bosses until the piston can be lifted off the connecting rod complete with rings. Make sure the piston is supported during this operation to avoid any risk of bending the connecting rod.

6   As each piston is removed mark it with the cylinder identification inside the skirt (left, right, centre). There is no necessity to mark the back and front of the piston because this is denoted by an arrow cast in the crown. This must face towards the front of the machine on reassembly.

7   Each piston is fitted with two piston rings. To remove the top ring spread the ends sufficiently with the thumbs to allow the ring to ease off.

8   With the lower ring you may need the assistance of a narrow blunt screwdriver for removal. Take great care in removing the piston rings as they are very brittle and can easily be broken.

8 Dismantling the engine/gearbox - removing the left hand engine covers

1   The left hand side of the engine is enclosed by the left cover and the front chain case cover. Under these covers are the engine sprocket, alternator and clutch release mechanism.

2   To disassemble the left hand side of the engine, remove the gearchange pedal (this is normally carried out prior to engine removal), followed by the front chain case cover.

3   To remove the engine cover undo the two securing crosshead screws and pull it off.

4.20c Or mounting studs

4.20d And then the mounting brackets

4.21 Lift the engine carefully from the right

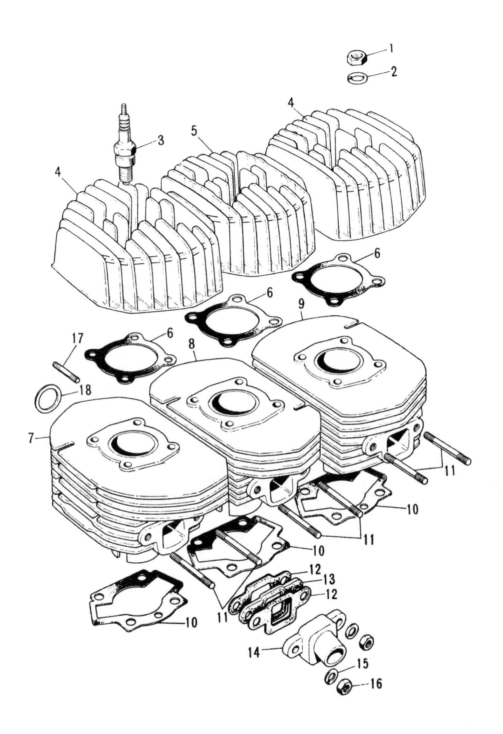

Fig. 1.1. Cylinder head and barrel assembly

1	Cylinder head nut - 12 off	5	Cylinder head - centre	10	Cylinder base gasket - 3 off
2	Spring washer - 12 off	6	Cylinder head gasket - 3 off	11	Stud - 6 off
3	Spark plug - 3 off	7	Cylinder, left-hand	12	Induction stub gasket - 6 off
4	Cylinder head, left and right hand - 2 off	8	Cylinder, centre	13	Heat insulator - 3 off
		9	Cylinder, right-hand		

14 Induction stub - 3 off
15 Spring washer - 6 off
16 Nut - 6 off
17 Stud - 6 off
18 Exhaust pipe gasket - 3 off

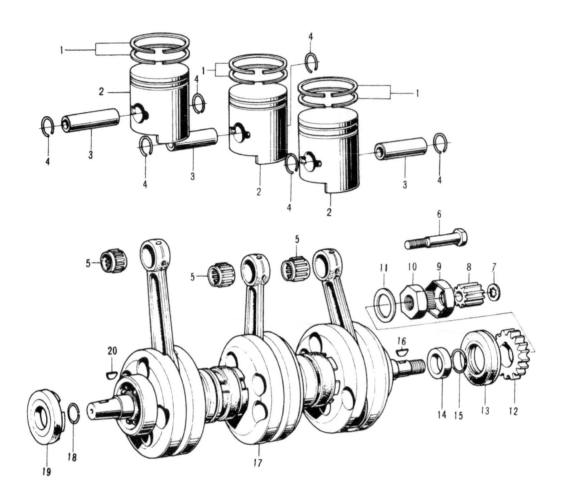

Fig. 1.2. Piston and crankshaft assembly

1	Piston ring set - 3 off	6	Oil pump pinion bolt	11 Lock washer
2	Piston - 3 off	7	Lock washer	12 Primary drive pinion
3	Gudgeon pin - 3 off	8	Oil pump pinion	13 Oil seal
4	Circlip - 6 off	9	Lock washer	14 Crankshaft collar
5	Small end bearing - 3 off	10	Nut	15 'O' ring

16 Woodruff key
17 Crankshaft assembly complete
18 Circlip
19 Oil seal
20 Woodruff key

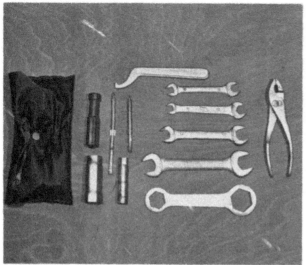

5.5 In addition to the tool roll

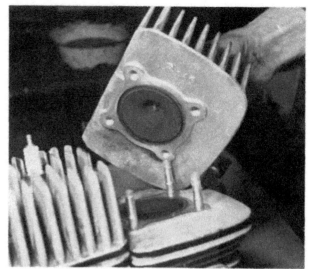

6.1 Slide the cylinder head up the holding down studs

6.2 The plugs can be left in situ, if desired

7.1 Lift them off the crankcase

7.2 Take care to support each piston

7.3 Stuff the crankcase mouth with clean cloths

7.4 Remove the circlips

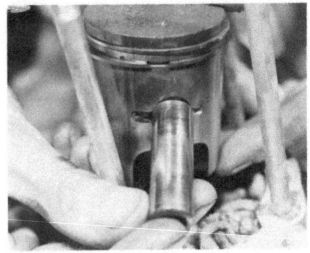
7.5 Push each gudgeon pin out of its piston boss

7.6 Arrow must face forwards when refitting

8.2a Remove the gear change pedal

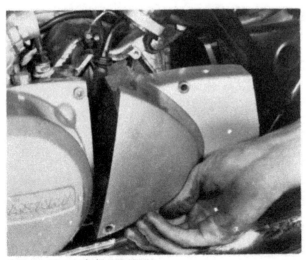
8.2b And front chain case cover

9 Dismantling the engine/gearbox - removing the engine sprocket

1   The engine sprocket is located under the left side engine
cover. It drives the rear wheel via the chain and will need cleaning
and careful examination. It is fabricated from wear resistant steel.
Note the road dirt on our 'fresh off the road' model.
2   To remove the sprocket, straighten the tab on the lockwasher
with a small cold chisel. It will have to be held firmly whilst the
nut and lockwasher is removed.
3   The method we adopted involved using a chain to hold the
sprocket against the case whilst the nut is removed. Use an old
chain for this purpose. After removing the nut and tab washer,
the sprocket can be withdrawn from the gearbox shaft. Great
care must be exercised if this system is used so that the case is
not damaged.

10 Dismantling the engine/gearbox - removing the alternator

1   To remove the alternator it is necessary to prevent the crankshaft from turning.
2   There are two ways of carrying this out. A long bar through
the connecting rod eyes, overlapping the side and resting on the
crankcase mouth will do the trick as will also a rolled up rag
placed between the primary drive and driven gears within the
right hand engine cover.
3   Identify the neutral switch on top of the case above the
engine sprocket and press down the terminal to remove the wire.
4   Remove the two long crosshead screws holding the stator and

lift it off.

5 Remove the timing cam bolt.

6 Pull off the rotor, using a sprocket puller. Avoid hitting the rotor if at all possible since this may destroy the magnetism.

7 Remove the key from the crankshaft and store it in a safe place.

11 Dismantling the engine/gearbox - removing the right hand cover and oil pump cover

1 The oil pump cover fits onto the right crankcase cover. This cover was removed prior to lifting the engine from the machine. See Section 4, paragraph 12 of this Chapter.

2 Mark the position of the kickstarter pedal, remove the pinch bolt and pull the lever from the splined shaft.

3 Ensure the transmission oil has previously been drained.

4 Remove the crosshead screws securing the right hand cover, remove the oil liners to the engine or the pump from its case, and lift off the cover. Place it on the bench for further stripping down.

5 Pull out very gently, using pliers, the tachometer gear.

6 Remove the mounting screws and lift out the oil pump.

7 Remove for inspection the tachometer gear mounting bracket, oil pump pinion, oil pump shaft and thrust washer.

8 It will be necessary to make a thorough check of the oil sealing in this compartment. This is due to the lubrication system of the machine. Oil is contained in the area between the right hand engine cover and the crankcase and is used for both lubricating and cooling the transmission, including the clutch.

9 Before removing the clutch, remove the mounting bolt and lockwasher from the oil pump pinion. Remove the oil pump pinion and lockwasher.

9.4 Remove the nut, tab washer and sprocket

10.3 The neutral switch shown was broken

10.4a Remove the two long crosshead screws

10.4b Lift off the stator

10.6 Using a sprocket puller to draw off the rotor

11.7 Remove the tachometer gear mounting bracket for inspection

11.9a Before removing the mounting bolt and lock washer

11.9b Remove the oil pump pinion

12   Dismantling the engine/gearbox - removing the clutch, clutch release and primary drive

1   The clutch fitted to the 'S' series Kawasaki machines is a wet multiplate type. It has six friction plates and five steel plates with steel rings fitted between the clutch steel plates and the friction plates. Endeavour to stow the items in the order of removal - this makes for easier assembly. The S3 model has 7 friction plates and 6 steel plates.
2   Remove the five crosshead bolts and take out the clutch springs, followed by the pressure plate, the plate lifter and the clutch pushrod.
3   The steel plates and friction plates that comprise the clutch assembly are fitted alternately, with a steel ring expander interposed between them.
4   Hold the clutch centre with a special tool (or by securing the drive sprocket end of the transmission with a chain) and remove the clutch centre mounting nut.
5   Remove the associated lock and flat washers, then withdraw the clutch centre, followed by the thrust washer.
6   Next remove the clutch housing from the driveshaft. Pull off the bush and the thrust washer.
7   The clutch release mechanism is removed by unscrewing the two crosshead mounting screws near the engine sprocket. Remove the last pushrod section at the same time.
8   The primary drive pinion can be removed before or after removing the clutch assembly.
9   Identify the primary pinion adjacent to the clutch area and straighten the lockwasher with a small cold chisel. Insert a

bar of suitable diameter through the connecting rod eyes to lock the crankshaft, then slacken and remove the nut on the primary drive pinion.
10 Remove the lockwasher and the primary gear pinion, then remove the Woodruff key from the shaft and stow it securely for reassembly later.

13   Dismantling the engine/gearbox - removing the external gearchange mechanism

1   The changing of gears on the 'S' series machine is initiated by an external mechanism turning a shift drum. Grooves on the shift drum cause selector forks in the gearbox to move and select the various gears.
2   The gearchange pedal assembly is on the left hand side of the machine. It consists of the pedal and an associated linkage assembly.
3   To disassemble the gearchange lever assembly, disengage the gearchange lever pawl from the shift drum pins and lift the gearchange lever assembly from the crankcase.
4   Remove the crosshead mounting bolt and lift out the stop lever and spring.
5   Unscrew and remove the retaining plate.
6   Examine the tension of the return spring and if it is weak or damaged, replace it before reassembly.
7
8   The stop lever spring is responsible with the stop lever for holding the drum in position. Replace it if it is weak or damaged.

12.2a Remove the five crosshead bolts

12.2b Take out the clutch springs

12.2c Then the pressure plate

12.2d The plate lifter

12.2e The push rod

12.3 The steel rings interposed between the clutch plates

12.4 Remove the clutch centre mounting nut

12.6a Remove the clutch housing

12.6b Pull off the bush

12.9 Insert a bar of suitable diameter to hold the crankshaft

13.3 Lift the gearchange lever assembly from the crankcase

13.4 Lift off the stop lever and spring

14 Dismantling the engine/gearbox - separating the crankcases

1   It is necessary to separate the crankcase to gain access to the crankshaft, transmission, kickstarter and gear shift components.
2   The crankcase separates into an upper and lower section. Both are diecast in aluminium alloy.
3   Remove the gearbox oil receiver.
4   Remove the crankcase oil receiver.
5   Ensure the clutch release mechanism has previously been removed.
6   Remove the nuts from the upper section of the crankcase and stow them in a small container.
7   Lightly tap the crankcase sections to separate them; there is no gasket between them, only good quality gasket cement used in small quantities.
8   Remove the two dowel pins that locate the crankcases with each other.
9   The crankcases will need cleaning after removing the gearbox components and the crankshaft assembly. Examine them for cracks or physical damage.
10 Each crank chamber is separate from the other, to maintain crankcase compression for each cylinder. This is accomplished by oil seals on the crankshaft assembly to make each chamber pressure-tight.
11 The transmission oil is prevented from entering by oil seals on the left end of the gearbox mainshaft and layshaft.
12 Oil leakage can be caused by a pressure build-up in the transmission housing due to the expansion of the oil. The breather hole, the breather, and its associated plastic pipe should be examined for correct functioning. It releases this pressure build-up, obviating the possibility of oil leakage between the crankcase and the left hand cover.
13 The upper crankcase has an oil passage for main bearing lubrication. This should be blown clear with an air line, if blocked.

15 Dismantling the engine/gearbox - removing the external kickstarter mechanism

1   The Kawasaki 'S' series machines are started by a kickstarter - they are not equipped with a starter motor. When the pedal is depressed the kickstarter shaft is operated. This shaft has a helical gear guide which meshes with the inner teeth of the kickstarter pinion. This pinion is moved along the shaft to mesh with the layshaft low gear. This in turn operates through the transmission linkage and turns the crankshaft.
2   Lift out the kickstarter assembly from the crankcase.
3   Remove the return spring guide and lift off the spring itself.
4   Remove the two circlips and the spring holder plate.
5   Remove the snap ring, the kickstarter pinion and kickstarter pinion holder off the shaft guide.
6   When reassembling, ensure there is not excessive play between the kickstarter pinion inner teeth and the kickstarter pinion on the guide shaft.
7   To check for smooth operation, operate the mechanism in both directions.

16 Dismantling the engine/gearbox - removing the gearbox components

1   The 'S' series models are fitted with a five speed, constant mesh, positive stop gearbox. Three selector forks move in the grooves of the gear shift drum within which they are located by a guide pin. As the shift drum turns the selector forks move in their slots, sliding gears into different meshing arrangements.
2   Commence disassembly by removing first the gearbox mainshaft from the upper crankcase which is now reversed in position during overhaul.
3   Next remove the layshaft, both shafts being removed with their respective gear clusters.
4   To remove the shift drum ensure the drum lever and the

13.6 Unscrew and remove the retaining plate

14.3 Remove the transmission oil receiver

14.4 Remove the crankcase oil receiver

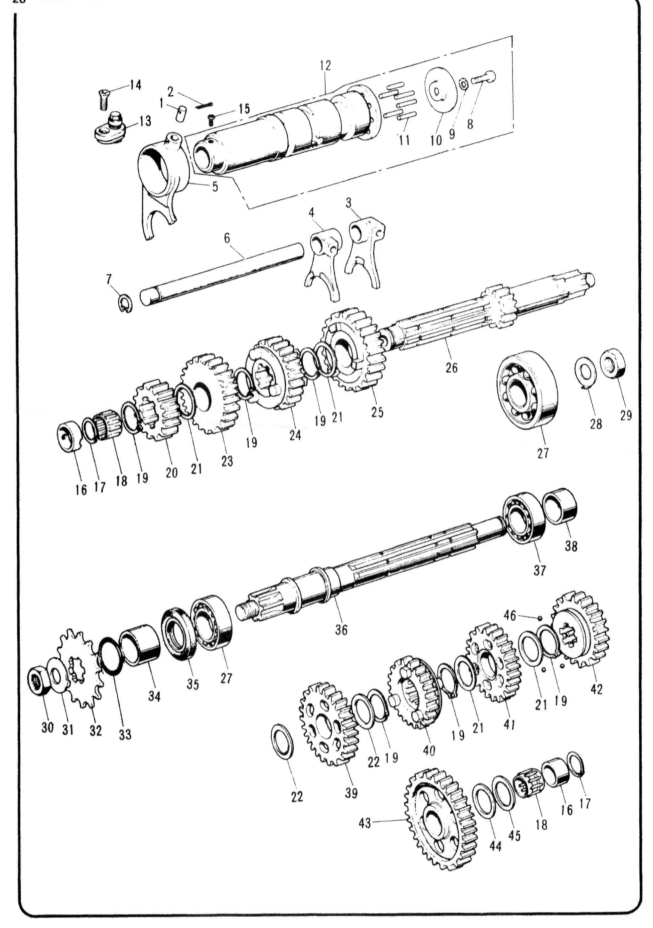

Fig. 1.3. Gearbox components

1   Selector fork guide pin
2   Split pin
3   Bottom gear selector fork
4   Second and third gear selector fork
5   Fourth and top gear selector fork
6   Selector fork spindle
7   Circlip
8   Screw
9   Lock washer
10 Gear selector drum plate
11 Gear selector drum pin - 6 off
12 Gear selector drum
13 Neutral indicator contact
14 Screw
15 Screw
16 Mainshaft bush - 2 off
17 Circlip - 2 off
18 Mainshaft needle roller bearing - 2 off
19 Circlip - 6 off
20 Mainshaft second gear pinion (19 teeth)
21 Thrust washer - 5 off
22 Thrust washer - 2 off
23 Mainshaft top gear pinion (27 teeth)
24 Mainshaft third gear pinion (23 teeth)
25 Mainshaft fourth gear pinion (25 teeth)
26 Mainshaft
27 Ball journal bearing - 2 off
28 Lock washer
29 Nut
30 Nut
31 Lock washer
32 Final drive sprocket (14 teeth, standard)
33 'O' ring, layshaft
34 Final drive sprocket collar
35 Oil seal
36 Layshaft
37 Ball bearing
38 Layshaft collar
39 Layshaft second gear pinion (34 teeth)
40 Layshaft top gear pinion (26 teeth)
41 Layshaft third gear pinion (31 teeth)
42 Layshaft fourth gear pinion (28 teeth)
43 Layshaft bottom gear pinion (40 teeth)
44 Thrust washer
45 Thrust washer
46 Ball bearing, 5/32 inch diam.- 3 off

15.1 The kickstarter gear is moved along the kickstarter shaft

15.2 Lift out the kickstarter assembly from the crankcase

15.3 And lift off the spring itself

15.4 Remove the two circlips and the spring holder plate

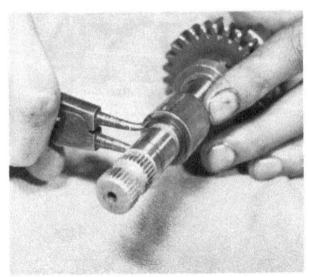

15.5 Remove the snap ring

15.6 Ensure there is not excessive play in the assembly

retaining plate have been removed.

5   Pull off the circlip on the selector rod, slide the rod out of position.

6   Lift off the two selector forks after removing the guide pins.

7   To free the shift drum, withdraw the split pin and the guide pin. Hold the remaining selector fork and pull out the shift drum from the crankcase.

17 Dismantling the engine/gearbox - removing the crankshaft

1   Using a mallet tap both ends of the crankshaft and lift it from the upper crankcase, complete with bearings and oil seals.

2   Remove, clean and examine the crankshaft bearing retaining rings in the upper crankcase.

18 Examination and renovation - general

1   Before examining the parts of the dismantled engine unit for wear it is essential that they are spotlessly clean. Clean all engine parts with a cleaning solvent as they are removed.

2   Use a solvent such as "Gunk" or "Jizer" for removing all traces of old oil and sludge that may have accumulated within the engine.

3   Examine the crankcase castings for cracks or other signs of damage. If a crack is discovered it will require professional repair.

4   Carefully examine each part to determine the extent of wear, checking with the tolerance figures given in the Specifications Section of this Chapter.

5   If you have any doubts take no chances and renew the item concerned.

6   Use a clean, lint-free cloth for cleaning and drying the items. This will obviate the risk of small particles obstructing the internal oilways, causing the lubrication system to fail.

19 Examination and renovation - crankshaft

1   If the crankshaft is damaged, or if the bearings have to be renewed a new or factory reconditioned unit must be obtained.

2   If you suspect the crankshaft is giving trouble and wish to replace it yourself, take the used one to a dealer and get an exchange. Check if you are within a 'warranty' period before you start work. The crankshaft is a very expensive item as it is a built-up assembly aligned to a high standard of accuracy.

3   The following are some of the items that can be checked with the crankshaft on the bench, provided it has previously given satisfactory performance.

4   Clean each of the main bearings with petrol. Lubricate them with oil and ensure there is no play. See that they rotate smoothly and have no trace of roughness.

5   There are four oil seals, one on each side of the crank chambers. These maintain chamber pressure differential. Inspect the oil seals for lip damage or for anything that might cause loss of compression. The centre oil seals are not replaceable items because they form an integral part of the built-up assembly.

6   Condensation within the engine, if the machine is used only infrequently, can cause failure of the big end bearings or the main bearings. When condensation troubles occur the flywheels become discoloured and rusty.

20 Examination and renovation - cylinder barrels

1   The usual indication of badly worn cylinder bores and pistons is excessive smoking from the exhausts and piston slap, a metallic rattle that occurs when there is little or no load on the engine.

2   If the top of the bore of each cylinder barrel is examined carefully it will be found that there is a ridge on the thrust side, the depth of which will vary according to the rate of wear that has taken place. This ridge marks the limit of travel of the uppermost piston ring.

3   Measure the bore diameter just below the ridge using an internal micrometer. Compare this reading with the diameter at the bottom of the cylinder bore which has not been subjected to wear. If the difference in reading exceeds 0.0012 inch (0.031 mm) the cylinder should be rebored and fitted with oversize piston and rings. Pistons are available in oversizes of 0.02 inch (0.50 mm) and 0.04 inch (1 mm).

4   Cylinder dimensions:

Model	Standard		Rebore at	
	Inch	mm	Inch	mm
S1	1.772	45	1.778	45.15
S2	2.086	53	1.093	53.15
S3	2.244	57	2.250	57.15

5   Ensure that the surfaces of the cylinder bores are free from score marks or other damage that may have resulted from an earlier engine seizure or a displaced gudgeon pin. A rebore will be necessary to remove any deep scores, irrespective of the amount of bore wear that has taken place, otherwise a compression leak will occur.

6   Remove all carbon from the exhaust ports so that the port area is completely unobstructed. If a build-up of carbon occurs, it will reduce the effective area of the port and may eventually give rise to back pressure problems.

21 Examination and renovation - pistons and piston rings

1   Attention to the pistons and piston rings can be overlooked if a rebore is necessary since new items will be fitted.

2   If a rebore is not necessary examine each piston closely.

3   Reject pistons that are scored or badly discoloured as the result of exhaust gases bypassing the rings.

4   Remove all carbon from the piston crowns using a blunt scraper, which will not damage the surface of the piston.

5   Check that the gudgeon pin bosses are not worn or the circlip grooves damaged.

6   Check that the piston ring grooves are not enlarged and that there is no build up of carbon on the inside surface of the rings or in the grooves of the pistons. Any build up should be removed by careful scraping.

7   Check that the ring pegs are intact and secure and that carbon build-up around this area has been removed.

An examination of the piston crowns will show if there has been a previous rebore; the amount oversize will be stamped on the crown. This information must be used when ordering a replacement.

8   To measure each ring for wear insert the ring about 0.2 inches into the cylinder using the piston to push it into position so that it is square with the bore.

9   Measure the end gap with a feeler gauge. The standard gap is 0.008 to 0.012 inch (2 - 3 mm). If the gap is greater than 0.031 inch (0.8 mm) replace the ring.

10 To check for wear in the piston ring grooves slip the outside surface of the ring into the groove and move the ring around the piston to ensure that there is no misalignment.

11 Temporarily fit the ring and with a feeler gauge measure the clearance between each ring and its groove at various points around the piston. These should be within the following limits:

Top ring
        Standard   Replace at
    0.0035 - 0.0051 inch (0.09 - 0.13 mm)   0.0067 in (0.17 mm)
Bottom ring
    0.0020 - 0.0035 inch (0.05 - 0.09 mm)   0.0047 in (0.12 mm)

12 To refit the rings fit the lower end first. Ensure any printing
on the ring is towards the top of the piston. Use the thumbs to
spread the rings just enough to slip on to the piston. Align the
end gaps with the locating peg in each ring groove.

16.2 Remove first the mainshaft

16.3 And then the layshaft

16.5a Pull off the circlip on the selector rod ...

16.5b ... and slide the rod out of position

16.7a Remove the guide pin and ...

22 Examination and renovation - cylinder heads

1   It is necessary to remove all carbon deposits from each cylinder head.

2   Ensure the spark plug has been removed and use a blunt headed scraper so that the combustion chamber is not damaged. The rounded head of an old hacksaw blade is a useful instrument for scraping the head. Do not damage the gasket surface area.

3   Examine the cylinder heads for linearity especially if traces of oil leakage at the head joint are evident.

4   Most cases of cylinder head misalignment are due to unequal tensioning of cylinder head nuts and bolts. The tightening torque should be 16 ft lbs (2.2 kg m) and the nuts and bolts tightened in the correct sequence. If the distortion is only slight, the jointing surface can be restored by rubbing down with emery cloth wrapped around a sheet of plate glass. Lay the emery-covered glass on the bench and rub down with a rotary motion, taking care that only a minimum amount of metal is removed in this manner. Check with a ruler or straight edge.

23 Examination and renovation - engine sprocket

1   Pay particular attention to the engine sprocket after thorough cleaning. A worn sprocket will cause excessive chain wear and a marked power loss. If you are at all doubtful about the sprocket due to wear, then renew it.

2   Check against the following table to see if the wear is excessive. Measurements are taken at root diameter:

No of teeth	Standard		Limit	
	Inch	mm	Inch	mm
14	2.41	61.2	2.38	60.4
15	2.59	65.8	2.56	65.0
16	2.80	71.2	2.77	70.4

3   When refitting the engine sprocket use a new tab washer and bend up with the cold chisel after the retaining nut has been tightened fully.

4   If possible, renew the gearbox and rear wheel sprocket at the same time, together with the chain. It is bad practice to run old and new parts together, especially since it will necessitate even earlier renewal on the next occasion.

24 Examination and renovation - gearbox components

1   It should not be necessary to dismantle either of the gear clusters unless damage has occurred or an item requires attention or replacement.

2   Our illustration shows how both clusters of the five speed box are assembled on their respective shafts.

3   It is imperative that the gears be assembled in exactly the correct sequence otherwise constant gear selection problems will occur. Our illustration shows an assembled mainshaft.

4   In order to eliminate the risk of misplacement make rough sketches as the clusters are dismantled.

5   Dismantle and reassemble one cluster at a time and in sequence. Lay the items on a bench or a sheet of paper in order of removal ready for reassembly.

6   Refer to the accompanying photographs which show both the mainshaft and layshaft in dismantled form. Do not lose the three ball bearings from the layshaft fourth gear. Mark all shims to ensure correct replacement.

7   Selector forks should be checked for play between their prongs and the gear grooves in which they sit. Insert a thickness gauge between each prong and its groove wall and check the limits as follows:

Standard gap 0.0020 - 0.0098 in (0.05 - 0.25 mm)
Limit   0.024 in (0.6 mm)

8   It is possible to remove rough edges and nicks from gear teeth with an oilstone if necessary.

16.7b ... pull out the shift drum

20.5 Ensure that the surface of the cylinder bores is free from score marks

21.6 Check that the piston ring grooves are not enlarged

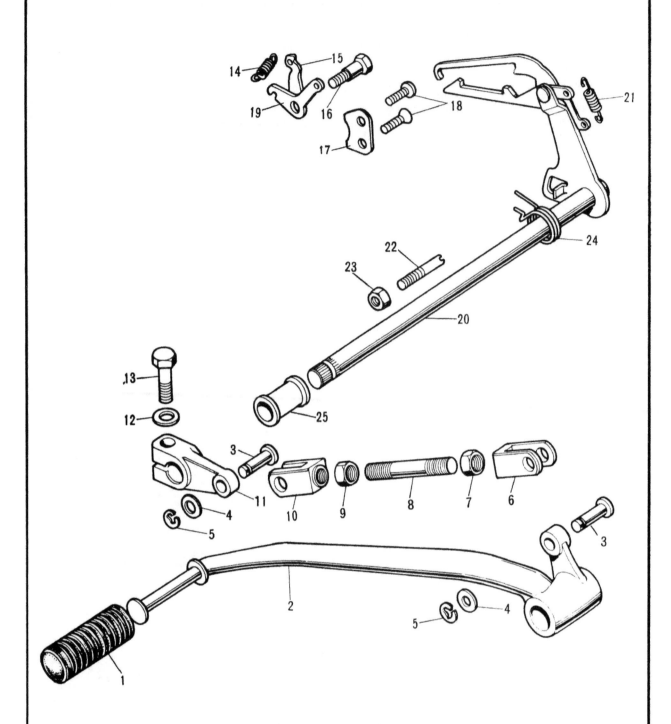

Fig. 1.4. Gearchange mechanism

1	Gear change lever rubber	8	Rod linkage
2	Gear change lever pedal	9	Nut
3	Clevis pin - 2 off	10	Joint (right-hand thread)
4	Plain washer - 2 off	11	Gear change lever link
5	Circlip	12	Spring washer
6	Joint	13	Bolt
7	Nut (left-hand thread)	14	Gear selector drum lever spring

15	Gear selector drum lever	20	Gear change lever assembly
16	Gear selector drum lever screw	21	Gear change lever return spring
17	Gear selector drum position plate	22	Gear change lever pedal return spring pin
18	Countersunk head screw - 2 off	23	Nut
19	Gear selector drum return lever assembly	24	Gear change lever pedal return spring
		25	Dust cover

9 Inspect the lips of the oil seals and renew any that are damaged.

10 Bearings should be checked for play and freedom of rotation, then lubricated, before reassembly.

11 When reassembling the layshaft fourth gear note the location holes for three steel ball bearings. Do not pack them with grease as they must be free to move.

12 Check the operation and condition of the needle roller bearings before installation. Do not omit to install all the circlips and shims of the needle roller bearings and gears.

13 Refit the bushes to the mainshaft and layshaft gear clusters prior to reassembly.

25 Examination and renovation - clutch, primary drive pinion and clutch release

1 The clutch springs must retain equal free length for the clutch to operate efficiently. The S1 model clutch springs should have a standard length of 1.358 inches and be rejected at 1.280 inches. The S2 and S3 clutch springs should have a standard length of 1.130 inches and be rejected at 1.051 inches.

2 The friction plates have cork linings. Check all friction plates for damage to this cork.

3 Check distance between friction plate projections and clutch housing indentations. These should be between 0.0020 and 0.0177 inches.

4 Measure the overall friction plate thickness. It should be about 0.118 inch. Replace at 0.106 inch.

5 The gear teeth of the clutch housing require careful checking for damage. If the damage is bad renew the clutch housing. Small nicks can be ground out using an oilstone. Ensure that there are no loose rivets in the clutch housing.

6 Ensure that the clutch housing bush and bearing are a good fit and undamaged. Too much play will cause clutch rattle.

7 Examine the clutch release gears and ensure that there is not too much play when they operate together. If in doubt replace both inner and outer gear which are only supplied as a unit.

8 Examine both pushrods for damage and linearity. Correct or renew the rods, depending on the extent of the wear or damage.

9 Examine the adjusting screw threads carefully and ensure the locknut can be tightened fully.

10 The primary drive pinion tooth surfaces should be examined for minor faults. These can be taken out with an oilstone. Major damage means a new replacement.

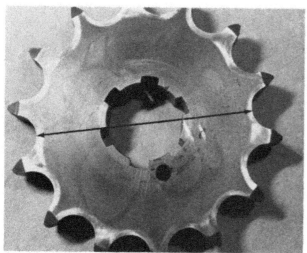

23.2 Measurements are taken at root diameter

24.2 The gear clusters assembled on their respective shafts

24.3 It is imperative that the gear pinions be assembled in exactly the correct sequence

24.6a Our illustration shows the mainshaft gear cluster dismantled

24.6b Followed by the layshaft

24.7a Forks should be checked for worn prongs

24.7b And the gear grooves in which they sit

24.8 It is possible to remove rough edges and nicks from gear teeth

24.9 Replace any damaged oil seals

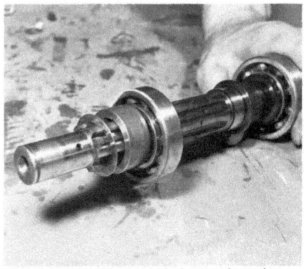

24.10 Bearings should be checked for freedom of operation

24.12a Check the operation and condition of the needle bearings

24.12b Reinstall all the circlips

24.13a Refit the bushes to the mainshaft

24.13b And to the layshaft

26 Reassembly of engine/gearbox - general

1   Before reassembly is commenced the various engine and gear-box components should be cleaned thoroughly and placed close to the working area.

2   Make sure old gaskets have been removed and that the mating surfaces are clean and undamaged. One of the better ways to re-move old gasket cement is to use methylated spirits. This will act as a solvent and obviate damage due to scraping.

3   Collect all necessary tools and have an oil can of fresh engine oil.

4   Have all new gaskets, oil seals and replacement parts available. Use the manufacturer's approved gaskets wherever possible.

5   Care should be taken to safeguard correct neutral selection when assembling the gear clusters. Operation of the gearchange pedal in either direction causes the shift drum to rotate between the drum pins one step.

6   Location of neutral, which is between first and second gears, is however a half step operation.

7   Three steel balls are fitted inside the output shaft fourth gear. Three grooves are cut into the layshaft fourth gear area. They are spaced around the shaft. The steel balls are kept out of the shaft

by inertia when the layshaft is turning. But if the shaft stops one or more of the balls will drop into a groove.

8   During the gearchange operation from low gear to neutral gear, fourth gear, due to fork movement, moves along the lay-shaft until the balls reach the end of the groove and halt the slide.

9   The groove is machined for only half the length of the fourth gear's movement thus the gear halts halfway. The selector fork movement is also stopped halfway by this control and in turn the change drum cannot turn a complete step. The net result is that gears do not mesh and stop in the neutral position.

27 Reassembly of engine/gearbox - replacing the crankshaft

1   Carefully insert the bearing retaining rings into the upper crankcase. There are three of these rings in the crankcase.

2   Align the groove in each ball bearing to its ring and seat the crankshaft in the serviced upper crankcase.

3   It is advisable to tap the bearings very lightly with a mallet to ensure that they are correctly located in the crankcase.

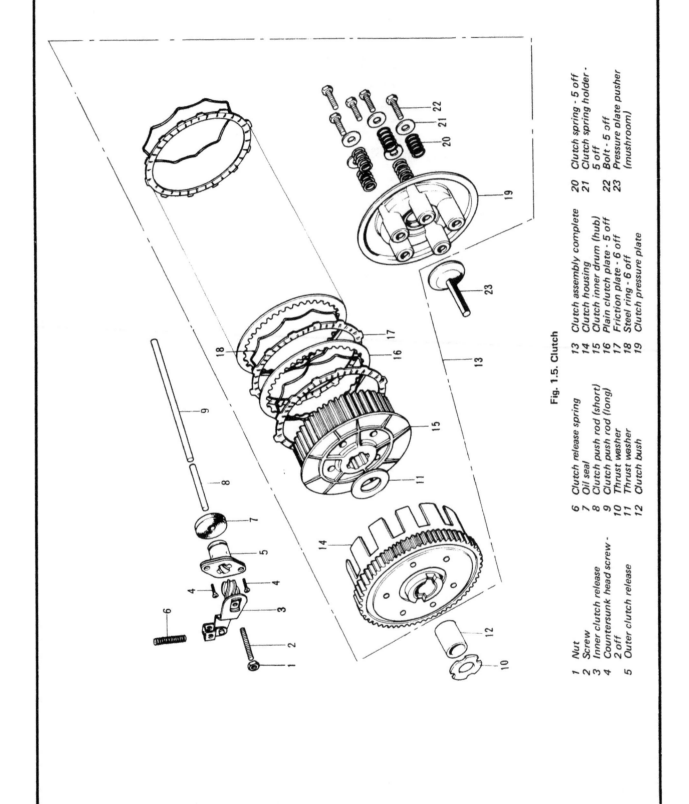

Fig. 1.5. Clutch

1 Nut
2 Screw
3 Inner clutch release
4 Countersunk head screw - 2 off
5 Outer clutch release

6 Clutch release spring
7 Oil seal
8 Clutch push rod (short)
9 Clutch push rod (long)
10 Thrust washer
11 Thrust washer
12 Clutch bush

13 Clutch assembly complete
14 Clutch housing
15 Clutch inner drum (hub)
16 Plain clutch plate - 5 off
17 Friction plate - 6 off
18 Steel ring - 6 off
19 Clutch pressure plate

20 Clutch spring - 5 off
21 Clutch spring holder - 5 off
22 Bolt - 5 off
23 Pressure plate pusher (mushroom)

28   Reassembly of engine/gearbox - rebuilding the gearbox assembly

1   Slide the shift drum into position over the 4th/5th gear selector fork.
2   Refit the guide pin and a new split pin to retain it.
3   At the end of the drum farthest from the selector mechanism you will need to have a pan headed screw in position. This will line up, when neutral is selected, with the neutral switch locating port. When the neutral switch is fitted and the cable is connected, the lighting circuit for the neutral lamp will be from the lamp to the neutral switch terminal on the crankcase. Contact is made via the spring copper contact of the neutral switch to earth through the pan headed screw when neutral is selected.
4   Slide the selector rod into position through the two selector forks and refit the guide pins and do not omit the circlip around the shaft.
4   Check the pins and guides for free movement.
5   Refit the gearbox bearing retaining rings into the crankcase and place the gear clusters and shafts into position. Check to ensure the needle roller bearings have located correctly.
6   Refit the oil receiver at the end of the layshaft and the crankshaft oil receiver.

29 Reassembly of engine/gearbox - replacing the kickstarter mechanism

1   Replace the kickstarter pinion and holder back on the kickstarter guide shaft.
2   Replace the snap ring.
3   Refit the spring holder plate with its associated two circlips. Attention should be paid to the angle it makes with the stop lever.
4   Replace the kickstarter spring and its guide.
5   Seat the complete assembly back into the crankcase and ensure that the kickstarter gear holder is located correctly into the crankcase.
6   The kickstarter stop is screwed in after crankcase assembly is complete.
7   When the kickstarter is released during engine starting the spring under tension will return the kickstarter and shaft to their starting points. The kickstarter stop lever hits the crankcase mounted stop, which prevents further travel.
8   To give sufficient spring tension turn the kickstarter stop counterclockwise 150° with the kickstarter attached and the crankcase reassembled.

30 Reassembly of engine/gearbox - joining the crankcases

1   Ensure all components are completely installed within the lower crankcase before reassembly.
2   Recheck that the crankcase oil receiver and the gearbox oil receiver have been replaced.
3   Apply gasket cement such as Hermetite Golden on the lower crankcase gasket surface. The gearbox components have already been installed in the upper crankcase.
4   Refit the oil drain plug gasket and then the oil drain plug. Tighten the plug.
5   Do not forget the dowel pins.
6   Refit the crankcase nuts. If the studs have also been removed, refit these first.
7   Tighten the crankcase nuts to the following torque settings:
12 x 6 mm nuts 10.8 - 11.6 ft lbs (1.5 - 1.6 m kg)
8 x 8 mm nuts   15.9 - 19.5 ft lbs (2.2 - 2.7 m kg)

8   Ensure the crankcase breather connection and breather hose are refitted.

31   Reassembly of engine/gearbox - replacing the external gearchange mechanism

1   The external gearchange mechanism is replaced in the reverse order of removal. Ensure the drum pins are in position and

26.7 Three steel balls are fitted inside the output shaft fourth gear

26.8 Three grooves are cut into the output shafts fourth gear operating area

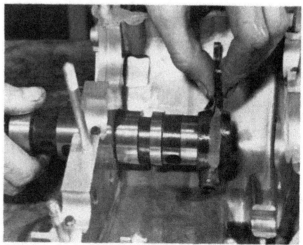

28.1 Slide the shift drum through the selector arm

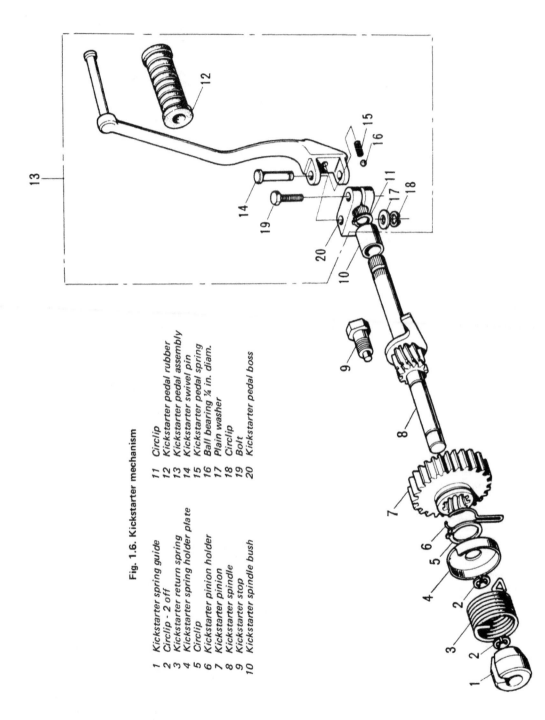

Fig. 1.6. Kickstarter mechanism

1 Kickstarter spring guide
2 Circlip - 2 off
3 Kickstarter return spring
4 Kickstarter spring holder plate
5 Circlip
6 Kickstarter pinion holder
7 Kickstarter pinion
8 Kickstarter spindle
9 Kickstarter stop
10 Kickstarter spindle bush

11 Circlip
12 Kickstarter pedal rubber
13 Kickstarter pedal assembly
14 Kickstarter swivel pin
15 Circlip
16 Ball bearing ¼ in. diam.
17 Plain washer
18 Circlip
19 Bolt
20 Kickstarter pedal boss

28.2 Refit guide pin and a new cotter pin

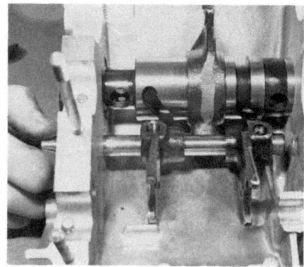
28.3a Slide the selector rod into position

28.3b Refit the guide pins and

28.3c Don't omit the circlip

28.4 Try the pins and guides for free movement

28.5 Refit the bearing retaining rings

28.6a Refit the oil reservoir, followed by ...

28.6b ... the crankcase oil receiver

29.4 Replace the kickstarter spring and its guide

29.5 Ensure that the kickstarter gear holder is located correctly

29.6 The kickstarter stop is screwed in after the full assembly

30.1 Ensure all components are completely installed

replace the side cover.

2   Refit the positioning plate and replace and tighten the mounting screws. Centre pop the crosshead mounting screws after tightening to ensure they cannot work loose.

3   Replace the stop lever and its spring and refit and tighten the bolt and screw.

4   Slide in the gearchange lever shaft. Check the location of the rubber cap and spacer at the lever end of the shaft.

5   Engage the lever with the pins of the shift drum.

6   Ensure the gearchange return spring is in position then fit the return spring pin.

7   Fit the locknut onto the pin and ensure that the locknut is tightened securely.

8   The gearchange lever mechanism is fitted after the engine has been reinstalled in the frame.

32 Reassembly of engine/gearbox - replacing the clutch, clutch release and primary drive pinion

1   Care must be taken when refitting the primary drive pinion to reinsert the key and to align the pinion with the key slot in the end of the crankshaft.

2   Fit a new lockwasher so that its projection aligns with the hole in the primary drive pinion.

3   Fit the primary drive pinion retaining nut and tighten securely. Note the method of preventing engine rotation whilst the nut is tightened in photo 32.3

4   Bend up one side of the lockwasher.

5   Remember to refit, in this order, the oil pump pinion lockwasher and then the bolt. These are on the same shaft. Tighten the bolt after reassembling the clutch. Use a bar to hold the crankshaft during this operation.

6   Refit the clutch release mechanism first ensuring that the oil seal has been renewed, or is fit for further use.

7   Refit in this order: oil seal, pushrod, bush, pushrods, inner and outer release gear, release gear securing corsshead screws, adjuster screw, locknut.

8   The adjuster screws will have to be re-aligned before using the machine.

9   The clutch must now be fitted. First fit the thrust washer last taken off and the bush.

10 Fit the clutch housing. Next the thrust washer, clutch centre, flat washer and lockwasher.

11 Refit the retaining nut. If the primary drive pinion is held fast by a bar through the connecting rod eyes and the clutch housing is meshed with it, the special service tool is not required. Tap over the tab washer to retain the nut.

12 Refit the steel plates, friction plates and steel rings into the housing. Do not forget the pressure plate lifter.

13 Next fit the pressure plate, spring guides, clutch springs and finally the five mounting bolts.

14 If the clutch, clutch release and primary drive components were first collected together in order in a tray this should facilitate reassembly in the reverse order of dismantling without problems.

33 Reassembly of engine/gearbox - replacing the right hand cover and oil pump cover

1   Place the right hand engine cover with a new gasket cemented in position on the bench for refitting the following items.

2   Refit the oil pump shaft, mounting bracket, nylon drive pinion, thrust washer and screws.

3   Refit the oil pump. The inlet and the outlet pipes must be refitted after the engine/gearbox is back in the machine.

4   The throttle cable and cable holder will be refitted after the engine/gearbox is back in the machine.

5   Examine the tachometer bush. It is fitted with an oil seal and an O ring. If the seal has been removed or damaged it must be renewed.

6   When replacing the tachometer bush, press it in as far as it will go. The lip section must on no account be damaged.

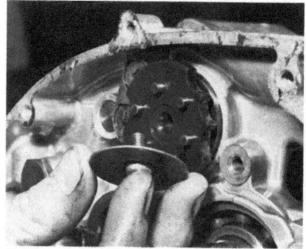
31.1 Ensure the drum pins are in position

31.2 Centre punch the crosshead mounting screws

31.3 Replace the stop lever and its spring

31.5 Engage the lever with the drum pins

32.1 Align the primary drive pinion with the keyway

32.2 Fit a new lock washer and the retaining nut

32.3 Tighten the primary drive pinion nut securely

32.5 Tighten the bolt after reassembling the clutch

32.8 The adjustment screw will have to be set

32.9 First fit the thrust washer and the bush

32.10 Then fit the clutch housing

32.11 Tighten the mounting nut with the engine locked

32.12 Don't forget the pressure plate lifter

32.13 Finally fit the pressure plate

33.2 Refit the oil pump shaft and mounting bracket etc. in the cover

44 Chapter 1/Engine, clutch and gearbox

7 At this stage examine the kickstarter oil seal. If damaged or removed it requires renewal.

8 Refit the right hand cover to the crankcase assembly and ensure the screws are correctly tightened.

9 Refill with SAE 10W30 or 10W40 transmission oil. Capacity is 1.1 litres or 0.97 Imp quarts.

10 Ensure the dipstick is not bent, then check dipstick oil reading.

34 Reassembly of engine/gearbox - refitting the alternator and engine sprocket

1 Refit the key into the keyway in the end of the crankshaft.

2 Insert the rotor and push it home on the crankshaft.

3 Refit the timing cam ensuring that the timing cam pin engages with the key inserted into the rotor.

4 Refit and tighten hexagon bolt and washer that retain the rotor on the crankshaft.

5 Refit the stator assembly over the shaft and refit and tighten the two long crosshead screws retaining it. Refit the alternator cable.

6 Examine the cam and contacts to ensure no damage has occurred during refitting. Ensure the felt oil pads are in position.

7 Refit the engine sprocket with a new tab washer. It will be necessary to secure the sprocket when tightening the sprocket nut.

8 Lightly bend over the tab washer into position on the nut.

9 Refit the neutral switch electrical connection.

10 Do not refit the left hand cover because it will be necessary to make clutch adjustments after the engine is reinstalled in the frame.

11 It will be necessary to reset the ignition timing before running the machine. Refer to Chapter 3, Section 7 for the correct procedure.

35 Reassembly of engine/gearbox - refitting the pistons and cylinder barrel

1 Ensure there is a clean rag in each crankcase mouth before re-placing the pistons. If a circlip is inadvertently dropped into the crankcase a further stripdown could be necessary.

2 Fit the pistons in their original order with the arrow embossed on each crown pointing towards the front of the machine. This is necessary because the gudgeon pin is slightly offset to prevent piston slap.

3 Tight fitting gudgeon pins can be fitted by heating the pistons.

4 The gudgeon pins and piston bosses should be oiled before fitting.

5 Use new circlips and thoroughly check their location within the groove inside each piston boss. Ensure the circlip opening does not face the piston groove.

6 Before fitting together the gudgeon pin, needle roller bearing and connecting rod check their radial play. Standard clearance is 0.00012 to 0.00086 inch (0.003 - 0.022 mm). Replace the bearing if this reaches 0.004 inch (0.10 mm).

7 Ensure that each piston ring end gap is aligned with the locating peg in the ring groove.

8 Lubricate the piston and cylinder and insert the piston into the lower end of the cylinder.

9 Compress each ring as it goes into the cylinder. There is a tapered lead-in at the base of each bore to aid this.

10 It is necessary to renew the three cylinder base gaskets if air and oil leaks are to be obviated. Make sure the base gasket is fitted so that the oil holes align correctly.

36 Reassembly of engine/gearbox - refitting the cylinder heads

1 Use new cylinder head gaskets when replacing the cylinder heads.

2 Slide the heads carefully back over the studs and tighten the retaining nuts by hand.

3 The nuts should then be tightened by a torque spanner to 16 ft lbs or 2.2 m kg. Tighten them in a diagonal sequence, to avoid distortion of the head.

37 Replacing the engine/gearbox unit in the frame

1 Although refitting the engine can be carried out single-handed the presence of an assistant to steady the frame and parts as the engine is lifted in is advisable.

2 Lift the complete engine unit into the frame from the right hand side and insert the lower rear engine bolt. The engine can now be manoeuvred to allow the three other engine bolts and associated frame mounting plates to be positioned correctly.

3 Replace the nuts and washers on the bolts where necessary and tighten them fully after correct positioning of all bolts and plates has been achieved.

4 A torque setting of 3 to 4 ft lbs is recommended for all four engine mounting bolts.

5 Remake the connections to the alternator wiring by rejoining the connectors.

6 Reconnect the clutch cable at the handlebars and at the engine. It is then necessary to adjust the clutch as follows:

7 At the engine do not completely tighten the clutch locknut and loosen the grooved screw until the clutch lever moves easily.

8 At the handlebars loosen the locknut and turn the adjuster to give ¼ inch thread length between the adjuster and locknut.

9 Identify the clutch cable adjuster between the handlebars and the release mechanism. Loosen the inbuilt locknut and turn the cable adjuster until the clutch release lever makes an 80° angle with the mounting screw. Now tighten this locknut.

10 At the engine turn in the lever adjustment screw until it becomes stiff. This indicates that the clutch is starting to operate. Hold the screw in this position with the screwdriver and tighten the locknut.

11 At the handlebar lever turn the adjuster to give 1/16 to 1/8 inch (2 to 3 mm) play. Tighten the milled locknut. Clutch adjustment is now correct.

12 Refit the final drive chain. Reconnection with the split link is made easier if the ends of the chain are pressed into the rear wheel sprocket after the chain has been threaded around the final drive sprocket of the gearbox.

13 When refitting the chain ensure that the closed end of the clip on the master chain link faces the direction of chain movement, ie forwards on the upper chain run.

14 Refit the front chain cover.

15 Refit the gearchange pedal to the marks previously inscribed and tighten the clamp bolt.

16 Remake the oil feed connections to and from the oil pump if not previously connected.

17 Refit the oil pump cable connection and ensure the cable holder crosshead screws are securely tightened. The throttle cable will need re-adjusting. Refer to Chapter 2, Section 15 for the procedure to be used.

18 Refit the carburettors using new O ring seals at their flange joints. Make sure the clamp screws are tight.

19 Reconnect the carburettor fuel pipes and the air inlet pipes.

20 Reconnect the tachometer cable and refit the oil pump cover.

21 Refit the three exhaust pipes and silencers. Attach the silencers loosely first, whilst the cylinder head joints are remade using a new sealing gasket in each case.

22 Slide up the exhaust pipe holders and refit the washers and nuts.

23 Refit the exhaust pipe clamps and tighten the crosshead screws securing them.

24 Lower the petrol tank into position again ensuring that no cables are pulled or trapped.

25 Remake the connections from the petrol tap to the carburettor float chambers. Check to ensure there is fuel in the tank.

26 If the battery has been removed, replace it in its holder.

27 Ensure the connections to the battery are perfectly clean with a smear of vaseline (mineral jelly). Note that grease is not a substitute as it insulates the terminals. Carefully tighten the terminals with the fingers only.

28 Refill the oil tank with the correct quantity of oil. The filler is located on the top of the crankcase and has an integral dipstick.

29 Double check the following before attempting to start the

34.2 Refit the key and insert the rotor

34.3 Refit the timing cam

34.4 Refit and tighten the hexagon bolt and washer

34.5 Refit the alternator cable

35.2 Refit with the arrow cast on each crown facing forwards

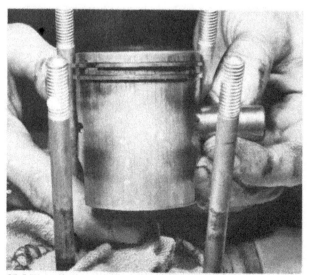
35.3 Tight fitting gudgeon pins can be installed by heating the piston

35.5 Use new circlips

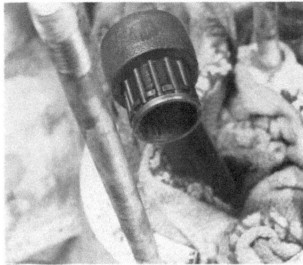
35.6 The needle roller bearing small end

37.2 Lift the complete unit into the frame from the right hand side

37.3 Locate all the mountings before tightening

37.7 Loosen the screw until the clutch lever moves easily

37.8 Have a ¼ inch thread length at the handlebar lever adjuster

37.9 Set the clutch release lever at an 80° angle with the mounting screw

37.11 Give 2.3 mm free play

37.14 Refit the chain cover

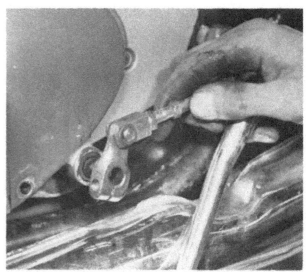

37.15 Refit the gear change ledal

37.21 Use a new sealing gasket

37.22 Refit the washers and nuts

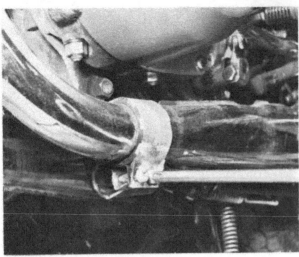

37.23 And tighten the crosshead screws

engine: engine oil and gearbox oil refilled (SAE 10W30 or 40), all nuts, bolts and screws reinserted and tightened. The following adjusted: idle speed, starter cable, oil pump cable, clutch cable, clutch, throttle, ignition correctly timed, brakes adjusted - a must!, brake light switches functioning, drive chain adjusted.
30 Check the operation of the oil pump as detailed in Chapter 2, Section 15.

38 Starting and running the rebuilt machine

1   Ensure adequate supplies of petrol and oil have been added to the machine.
2   Switch on the ignition with the key.
3   Switch to ON at the petrol tap - wait a few seconds.
4   Operate the START lever on the handlebars and depress the kickstarter.
5   As soon as the machine starts to run evenly keep it running at low speed to allow oil pressure to build up and the oil to circulate.

6   Exhaust smoke may appear from the exhausts initially due to the amount of oil used when assembling the engine. This should gradually disappear as the engine settles down.
7   Check the exterior of the machine for oil leaks or blowing gaskets.
8   Check the gear selection for smooth and positive operation.
9   Check the operation of the brakes both front and back, also the brake light.
10 Check the lights and horn and ensure the machine is fully roadworthy.

39 Taking the rebuilt machine on the road

1   Any rebuilt machine will need time to settle down even if parts have been replaced in their original order. Treat the machine gently for the initial few miles to ensure free circulation of oil through the lubrication system. This gentle treatment will also assist any newly fitted parts to bed down.
2   If the engine has been rebored the engine will have to be run in again, as if it were new. This includes skilful use of the gearbox and less throttle until about 500 miles have been covered. Keep the machine as lightly loaded as possible and even after the initial 500 miles only increase the load on the engine gradually. Watch the pillion load.
3   If you have had to fit a new crankshaft only the recommendations above can be lessened to some extent but a reasonable run-in period is advised.
4   Keep an eagle eye on lubrication during this period. If in doubt, stop and investigate as to run without oil will damage your newly repaired engine.
5   Take similar precautions if you hear peculiar noises.
6   It is advisable to retorque the cylinder heads down after the 500 mile period. These should be at 16 ft lbs (2.2 m kg).
7   With your machine ready for many miles of travel by day or night PLEASE wear something bright on the road. A full range of protective clothing in lime green and orange has been introduced by Kawasaki as part of their policy of brightening up the motorcycle scene and to get away from the old black jacket image. The new gear was designed in USA specifically to cope with the problem of car drivers failing to notice motorcycles in potential accident situations.

40 Fault diagnosis - engine

Symptom	Cause	Remedy
Engine will not start	Spark plugs defective	Remove the plugs one by one and lay them on cylinder head. Check if spark occurs with ignition ON and engine rotated.
	Dirty or closed contact breaker points	Examine condition of points and check gap.
	Faulty or disconnected condenser	Check whether points arc when separated.
	Flat battery	Remove and recharge battery.
Engine runs unevenly	Ignition and/or fuel system fault	Check each system independently as though engine will not start.
	Blowing cylinder head gasket	Leak should be evident from oil leakage where gas escapes.
	Incorrect ignition timing	Check accuracy and if necessary reset.
Lack of power	Fault in fuel system or incorrect ignition timing	See above.
Heavy oil consumption	Cylinder barrels in need of rebore	Check for bore wear, rebore and fit oversize pistons if required.
	Damaged oil seals	Check engine for oil leaks.
Excessive mechanical noise	Worn cylinder barrels (piston slap)	Rebore and fit oversize pistons.
	Worn big end bearings (knock)	Replace crankshaft assembly.
	Worn main bearings (rumble)	Fit new journal bearings and seals. Replace crankshaft assembly if centre bearings are worn.

Engine overheats and fades	Lubrication failure	Stop engine and ensure oil is getting to internals. Check oil level in crankcase.
	Wrong ignition timing	Check and re-set if necessary.
	Wrong spark plugs	Use recommended grades.

41 Fault diagnosis - gearbox

Symptom	Cause	Remedy
Difficulty in engaging gears	Selector forks bent	Renew.
	Gear clusters not assembled correctly	Check gear cluster arrangement and position of thrust washers.
Machine jumps out of gear	Worn dogs on ends of gear pinions	Renew worn pinions.
	Stopper arms not seating correctly	Check stopper arm action
Gearchange lever does not return to original position	Broken return spring	Renew spring.
Kickstarter does not return when engine is turned over or started	Broken or poorly tensioned return spring	Renew spring or retension.
Slipping kickstarter	Ratchet assembly worn	Parts must be renewed.

42 Fault diagnosis - clutch

Symptom	Cause	Remedy
Engine speed increase as shown by tachometer but machine does not respond	Clutch slip	Check clutch adjustment for free play at handlebar lever. Check thickness of inserted plates.
Clutch operation stiff	Damaged, trapped or frayed control cable	Check and replace if necessary. Make sure cable is lubricated and has no sharp bends.
	Bent operating pushrod	Check the pushrod for trueness.
Difficulty in engaging gears, gear changes jerky and machine creeps forward when clutch is withdrawn. Difficulty in selecting neutral	Clutch drag	Adjustment for too much free play. Check clutch drums for indentations in slots and clutch plates for burrs on tongues. Dress with file if damage not too great.

Chapter 2 Fuel system and lubrication

Contents

Specifications

Fuel tank capacity ... ... 3.08 Imp gallons (14 litres/3.70 US gallons)

Engine oil: 2 stroke oil ... ... 1.32 Imp quarts (1.5 litres/1.60 US quarts)

Gearbox oil: SAE 10W30 or SAE 10W40 ... ... 0.97 Imp quarts (1.1 litres/1.16 US quarts)

Carburettor
Type ... ... Mikuni

Model	Type	Main jet	Needle jet	Jet needle	Pilot jet	Cutaway	Air screw	Fuel level
S1	VM22SC	75	0-2	4EJ9-3	17.5	2.5	1¼ turns out	1.1 ± 0.04 in 28 ± 1 mm
S2 (Eng. no.	VM24SC	85	0-2/4	4EJ4-3	25	2.0	1½ " "	1.06 ± 0.04 in 27 ± 1 mm
S2 0–3773)	VM24SC	82.5	0-2/4	4EJ4-3	25	2.5	1½ " "	ditto
S3	VM26SC	85	0-2/8	4EJ4-3	22.5	2.0	1¾ " "	ditto

1 General description

The fuel system consists of a fuel tank from which petrol is fed by gravity to three carburettors via a petrol tap. This is a manually operated tap and includes a built in filter.

All models are fitted with three Mikuni carburettors which have integral float chambers and manually operated chokes.

A large capacity air cleaner serves the dual purpose of supplying clean air to the three carburettors and cutting down the air intake noise.

Engine lubrication on 'S' models is by the Superlube system. Two-stroke engine oil is carried in a 1.5 litre tank. An engine driven oil pump supplies oil to the requisite areas of the engine. Engine speed and also throttle control ensures that the engine gets the correct amount of oil for the existing engine load.

The gearbox components of the machine are lubricated by oil in the crankcase which is separate from the engine lubricating system. 10W30 or 40 oil (1.1 litres) is carried in the right hand crankcase cover. A filler cap and dipstick fit into the top of this cover.

2 Petrol tank - removal and replacement

1 It is not necessary to remove the petrol tank for minor checks as it does not restrict work on the engine. The tank itself is made of corrosion resistant steel.

2 If the tank is removed with the petrol tap fitted, switch the lever to the left to the STOP position. This is for temporary work when the tank is kept outside the shed or garage.

3 If a lengthy job entails removing the tank, leave the three carburettor petrol feed pipes in position on the tap and push them into the neck of a petrol can capable of containing the petrol remaining. Switch the lever to the right to the RESERVE position and drain the fuel. Switch to STOP when the fuel has drained out.

4 Undo the strap securing the tank at the rear and lift out the tank from its two rubber supports at the front.

5 Refitting is the reverse procedure. Ensure the securing strap is refitted and reconnect the feed pipes to the petrol tap.

6 The filler cap is the snap action quick release type. There is no lock fitted on the S1 and S2 series tanks and in some areas it

might be advisable to arrange for your dealer to fit a lockable type cap as an extra.

3 Petrol tap - removal, dismantling and replacement

1   The petrol tap has three positions:

    To the left — STOP — S
    To the centre — ON — O
    To the right — RESERVE — R

2   In normal operation the fuel will flow whilst its level is above the top of the main pipe protruding into the tank. When switching to RESERVE fuel remaining in the tank can be gravity fed from the top of a filter at the base of the tank. The amount of fuel available in RESERVE is 2 litres or ½ US gallon or 1.76 Imp quarts.
3   To clean the tap, drain petrol as instructed in Section 2. Remove the tap by releasing the nut securing it to tank (using a set spanner across the flats). Clean out any sediment or rust in the tap or the bowl. Blow through with an airline or foot pump and check for leaks after refitting.
4   If the model has an automatic petrol tap a different procedure is necessary. The tap will have a diaphragm attached to it. In addition to the operations above if there are signs of leaking from the tap, remove the diaphragm from its cover. Clean the valve and seat. When reassembling ensure the vent holes re-align. Ensure no air leaks in the negative pressure tube linking the tap to the carburettor. Run the engine and check the operation of the automatic feed to ensure there is a full flow to each carburettor.

4 Carburettors - general

1   The 'S' series Kawasakis are each fitted with three Mikuni carburettors. The type of carburettor can vary between models and our procedures in this Chapter are based on the one removed from the S2 model. Each carburettor comprises:
A main jet and slide system — used for petrol supply during operation at high and medium speeds.
A pilot jet system — used for low speed operation.
A float mechanism — for maintaining the fuel level in the float chamber.
A starter system — to supply a rich fuel during the start operation.
2   A conventional throttle slide and needle arrangement is coupled with a main jet to control the amount of petrol/air mix fed to the engine. The main jet flow is controlled by the pressure drop in the air stream as it passes through the venturi.
3   The right and centre carburettors are identical but the left is assembled slightly differently to cater for its location in the machine. It is however identical in operation.
4   Each carburettor has a short tube extension adjacent to the pilot jet adjustment screw, for linking to the petrol supply.
5   Each carburettor has an overflow pipe adjacent to the petrol feed pipe. This has a plastic tube attached and will operate if the float needle valve sticks.
6   The cold start system is manually operated by cable linkage from the START lever. When the lever is depressed it raises a plunger in the carburettor body and permits a very rich mixture of fuel and air to be mixed with air drawn through the starter air inlet. The resultant very rich mixture comes from the starter outlet on the engine side of the throttle valve and is fed to the engine.

5 Carburettors - removal

1   The three carburettors may be disconnected with or without engine removal. They can also be removed as individual units.
2   Ensure all the pipes from fuel tap have been disconnected.
3   Loosen the clamp screws of the air inlet hose to each carburettor and pull off the tubes.
4   Loosen the screws in the retaining clips around the inlet mani-

folds.
5   Lift off the carburettor assembly.
6   Retain the plastic vent tube on the carburettor body.

2.4 Lift out the tank from its rubber supports at the front

3.1 The petrol tap has three positions

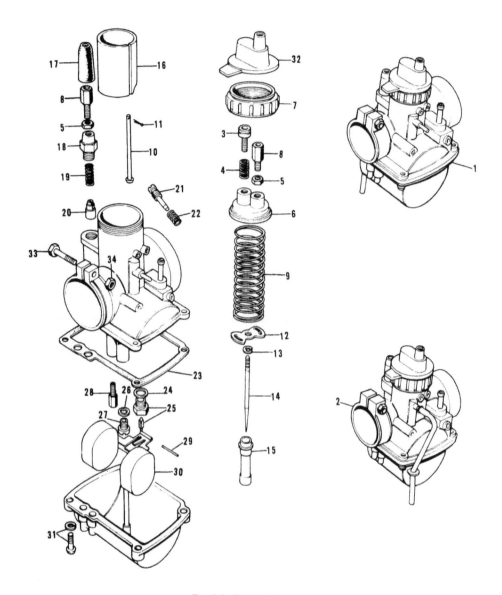

Fig. 2.1. Carburettor

1	Carburettor assembly, left-hand	9	Throttle valve return spring - 3 off	17	Rubber cap - 3 off	25	Float valve assembly - 3 off
2	Carburettor assembly, right-hand and centre	10	Throttle valve stop rod - 3 off	18	Starter plunger cap - 3 off	26	Main jet washer - 3 off
3	Throttle adjuster - 3 off	11	Split pin - 3 off	19	Starter plunger spring - 3 off	27	Main jet - 3 off
4	Throttle adjuster spring - 3 off	12	Throttle valve spring seat - 3 off	20	Starter plunger - 3 off	28	Pilot jet - 3 off
5	Cable adjuster lock nut - 3 off	13	Circlip - 3 off	21	Pilot jet adjusting screw - 3 off	29	Float pin - 3 off
6	Mixing chamber top - 3 off	14	Needle - 3 off	22	Pilot jet adjusting screw spring - 3 off	30	Float assembly - 3 off
7	Mixing chamber cap - 3 off	15	Needle jet - 3 off	23	Float chamber gasket - 3 off	31	Float chamber retaining screw - 12 off
8	Cable adjuster - 3 off	16	Throttle valve - 3 off	24	Float valve seat washer - 3 off	32	Rubber cap - 3 off
						33	Clamp screw - 3 off
						34	Nut - 3 off

6 Carburettor - dismantling, renovation and reassembly

1   Unscrew the start cable assembly from the side of the carburettor. Note that the spring is under tension.
2   Attached to this for removal are the rubber cap, the cable adjuster, locknut, plunger cap, plunger spring and the starter plunger itself.
3   Unscrew the mixing chamber cap and lift out the throttle slide assembly.
4   The carburettor slide, return spring, needle assembly and mixing chamber tap is attached to the throttle cable.
5   The throttle cable can now be released from the slide mechanism by removing a clip.
6   The jet needle can also be released from the throttle slide. Note its position first.
7   Invert the carburettor and remove the four retaining screws and lift off the float chamber. Note the fitted gasket. It need not be disturbed unless damaged.
8   Carefully remove the pivot shaft and lift out the float assembly.
9   The main jet is located in the centre of the mixing chamber housing and is unscrewed from the base. It is threaded into the base of the needle jet.
10   Remove the float valve and then the valve assembly and seating washer.
12   Remove the throttle pilot jet screw.
13   Check the rubber inlet tubes to the carburettors from the air cleaner and renew them if damaged.
14   Examine each of the carburettor components carefully and if in doubt renew the item concerned.
15   Ensure none of the floats is damaged. A small puncture can be soldered but use only a minimal amount of solder or the weight will affect float action.
16   The needle jet can be lifted out from the top of the unit when the main jet is removed. Check for damage and if in doubt renew.
17   Examine the throttle slide for wear. Worn slides give rise to air leaks and should be renewed.
18   The throttle needle is normally suspended by a circlip from the centre of the throttle slide. In some carburettors, they may be found in grooves other than the centre for mixture strength adjustment.
19   The start system is manually operated by cable linkage from the START lever. The plunger in the carburettor body should be checked for correct seating.
20   Each carburettor component should be cleaned extremely carefully. Compressed air is best and take great care. Even with a lint-free cloth fine particles can obstruct the internal air passages or jet orifices. Never use wire as it will enlarge the jet. Never use compressed air on an assembled carburettor as the float and float valve could be damaged.
21   Check the overflow port to ensure there is no obstruction and examine its associated plastic tube for serviceability.
22   Reassemble the carburettor components in the reverse order to the stripdown.

7 Carburettor - adjustment

1   Carburettor adjustment is not normally required unless you are replacing jets or the unit has recently been overhauled.
2   If you have reason to be dissatisfied with carburettor performance consult your main dealer. This performance could be affected by high altitudes or temperature extremes.
3   It is possible to have a different main jet, needle jet, jet needle, clip and throttle valve. Remember you will have to change three carburettor sets.
4   If you adjust the position of your needle jet on one carburettor the operation must be repeated on the other two.

8 Synchronising the carburettors

1   Power output will be unbalanced unless all three carburettors are correctly synchronised.

5.3 Loosen the screws to the air inlet to each carburettor

5.5 Lift off the carburettor assembly

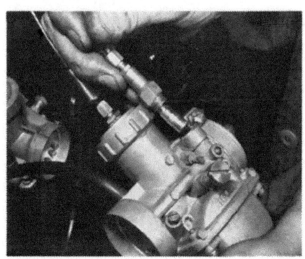
6.1 Unscrew the starter cable assembly

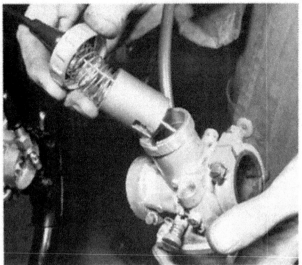

6.3 Lift out the throttle slide assembly

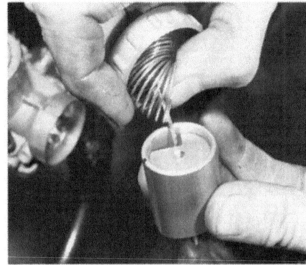

6.5 The throttle cable can now be released from the slide mechanism

6.7 Lift off the float chamber

6.7a Note the fitted gasket

6.9 The main jet is located in the centre of the mixing chamber housing

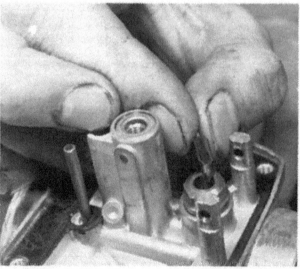

6.10 Remove the float valve

6.10a And then the valve assembly

6.11 Remove the throttle stop screw

6.12 Remove the pilot jet screw

2   Remove the carburettor air inlet hoses to check the operation of the throttle valves to ensure that the three carburettor slides enter the bore of the carburettor at the same time. Open the throttle to raise both slides to their maximum height and slowly close the throttle to check the entry of the slides into the bore.

3   A throttle cable goes to each carburettor. The three cables in turn are linked to one control cable which is operated by the throttle, via a junction box.

4   Identify the throttle cable locknut and make slack in the cable at the throttle grip by turning the cable adjuster.

5   Turn each of the three throttle stop screws counterclockwise until the throttle valves are fully closed.

6   You will find that some carburettors have the throttle stop screw on top (S1) whilst others have it fitted to the side (S2).

7   On S2 models adjust the cable guide on the top of the carburettor for zero play by means of the cable adjuster at that point. Loosen the locknut before and tighten after the adjustment. Move the cable guide up and down whilst making the adjustment until no play is felt.

8   Repeat the adjustment at each carburettor.

9   Identify the carburettor air screw.

10 Screw in each air screw until it is lightly seated, then turn back 1¼ turns (S1), 1½ turns (S2), 1¾ turns (S3).

11 Refit the air filter hoses and run the engine until it is warm, then at the following rpm (S1, S2 1300 - 1500 and S3 1100 - 1200) adjust each of the three throttle stop screws equally to achieve smooth idling at lowest engine rpm in that spread of rpm.

12 Go back to the throttle cable adjuster and make the required amount of slack as in paragraph 4 of this Section.

13 Loosen the locknut and adjust to provide 0.08 to 0.12 inch (2 - 3 mm) play at the twist grip. Tighten securely.

14 The starter cable will now need adjustment. It is the lever marked 'S' adjacent to the throttle grip. The starter cable adjuster is adjacent to the throttle adjuster but further inboard.

15 Loosen the locknut and turn the adjuster to create some play in the starter lever.

16 The control cable has three take-offs, one to each carburettor, via a junction box.

17 Identify the starter plunger control at each carburettor.

18 Adjust the outer sleeve of this control to give 0.04 - 0.08 in (1 - 2 mm) play on each carburettor by turning the adjuster whilst the locknut has been loosened. Move the cable up and down whilst adjusting until only slight play is felt.

19 Tighten the locknuts on each carburettor starter control cable after adjusting each in turn.

20 Go back to starter lever and re-adjust for about 1/16 - 1/8 in (2 - 3 mm) play.

21 Retighten the locknut on the starter cable adjuster.

22 The starter lever is used with a cold engine to create a rich mixture. Operation of the start lever lifts up the plungers adjusted in operation 19.

23 Turning the engine by kickstarter sucks petrol through a starter jet. The petrol/air mix from the starter system is in addition to the petrol/air mix from the pilot system with which it is in turn mixed.

9 Carburettor settings

1   Some of the carburettor settings, such as the sizes of the needle jets, main jets, needle positions etc, are predetermined by the manufacturer. Under normal circumstances it is unlikely that these settings will require modification, even though there is provision made. If a change appears necessary it can often be attributed to a developing engine fault.

2   As an approximate guide the pilot jet setting controls engine speeds up to 1/8 throttle. The throttle slide cutaway controls engine speed from 1/8 to 1/4 throttle and the position of the needle in the slide from 1/4 to 3/4 throttle. The size of the main jet is responsible for engine speed at the final 3/4 to full throttle. It should be noted that these are only guide lines. There is no clear demarcation due to degrees of overlap which occur between

the carburettor components.

3 Always err slightly on the side of a rich mixture - a weak mixture will overheat the engine. Reference to Chapter 3 will show how the condition of the spark plugs can be used as a guide to carburettor mixture condition.

10 Exhaust system - cleaning

1 The cleaning of the exhaust system should not be neglected as it has quite a pronounced effect on the performance of a two-stroke. The exhaust system must be inspected and cleaned out at regular intervals. The exhaust gases from a two-stroke are of a very oily nature which encourages sludge build-up. This build-up causes back pressures and affects engine breathing.
2 Cleaning is made easy by fitting the silencers with detachable baffles, held in position with bolts.
3 To remove the baffle tube remove the bolt at the back of the silencer and pull out the baffle tube.
4 A wash with a grease and oil solvent such as "Gunk" or "Jizer" and brushing with a wire brush will remove most of the carbon deposits but it is good practice to use a blow lamp to get off the residue. Keep the blow lamp operation away from petrol vapour.
5 Renew the rubber connector between each exhaust pipe and silencer if a leak develops.
6 At less frequent intervals such as on a decoke session the exhaust pipes themselves can be cleaned out.
7 You are advised not to replace your silencers with a much advertised alternative. A difference in 'note' may be very deceiving as to ultimate performance.

11 Air cleaner - dismantling, servicing and reassembling

1 The role of the air cleaner is to prevent road dirt from gaining access to the engine and causing wear to the cylinders, pistons and associated rings. A clogged air cleaner therefore can reduce engine power and performance.
2 Remove the left hand side cover and loosen the thumb screw at the top of the filter.
3 It is possible to ease out the air cleaner assembly through the left side of the frame avoiding the electrical components and connectors in that compartment.
4 When it is necessary to remove the air cleaner assembly baseplate the three carburettor hose clamps must first be removed to pull off the air inlet ducts.
5 Remove the three baseplate mounting screws.
6 Remove its two mounting screws to make the element available for cleaning.
7 Clean the element by air jet or immersing in petrol. If the element is badly choked it is advisable to renew it. Keep oil away from the element. Dry paper elements must be cleaned only by blowing with compressed air.
8 Note the arrow and the UP sign when refitting.

12 Engine lubrication

1 The 'S' series models carry two different oil supplies. One is for the engine lubrication and the other for gearbox lubrication. The engine oil is contained in a fibreglass tank below the seat on the right hand side of the machine. The tank holds 1.32 quarts (1.5 litres) and is fitted with the normal screw type filler cap with a gauze filter in the cap.
2 To gain access for servicing remove the cover and the securing bracket after draining the oil and removing the oil pump feed connection.
3 Kawasaki triple cylinder 'S' series machines use the Superlube system of oil injection. An engine driven oil pump supplies the engine lubrication. The pump output is controlled by the speed of the engine and also by the throttle position to give the correct amount of oil at any given engine speed.
4 The tanks containing fuel or oil should be cleaned out after long periods of use to get rid of any sediment developing. In any

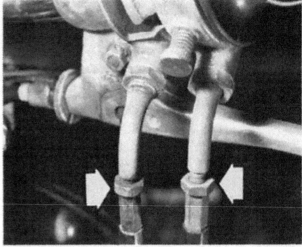
8.4 Identify the throttle cable adjuster locknut

8.5 Turn each of the three idle speed screws counter clockwise

8.16 It is the lever marked 'S' adjacent to the throttle grip

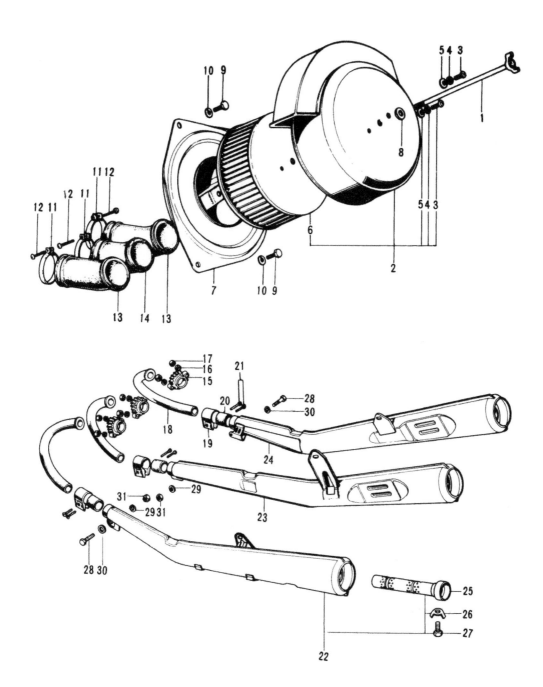

Fig. 2.2. Air cleaner and exhaust system

1 Air cleaner fitting
   bolt
2 Air cleaner assembly
3 Screw - 2 off
4 Spring washer - 2 off
5 Plain washer - 2 off
6 Air cleaner element
7 Air cleaner body
8 Plain washer

9 Bolt - 3 off
10 Spring washer - 3 off
11 Air inlet hose clamp -
   3 off
12 Screw - 3 off
13 Air inlet hose, left-hand
   and right-hand
14 Air inlet hose, centre
15 Exhaust pipe clamp - 3 off

16 Spring washer - 6 off
17 Nut - 6 off
18 Exhaust pipe - 3 off
19 Silencer connector - 3 off
20 Silencer connector holder -
   3 off
21 Screw - 6 off
22 Silencer assembly, left-
   hand

23 Silencer assembly, centre
24 Silencer assembly, right-
   hand
25 Baffle tube - 3 off
26 Lock washer - 3 off
27 Bolt - 3 off
28 Bolt - 2 off
29 Spring washer - 2 off
30 Plain washer - 2 off

Fig. 2.3. Oil pump

1 Oil pump assembly
2 Banjo union bolt - 3 off
3 Banjo union bolt
4 Banjo union connector
5 Oil pump washer - 8 off
6 V ring, oil pump camshaft
7 'O' ring, oil pump spindle
8 Oil seal, oil pump spindle
9 'O' ring, oil pump spindle
10 'O' ring, oil pump lower cap
11 Oil pump cable lever
12 Oil pump gasket
13 Screw
14 Screw - 2 off
15 Oil pump washer - 2 off
16 Left-hand oil pipe assembly
17 Centre oil pipe assembly
18 Right-hand oil pipe assembly
19 Banjo union bolt - 3 off
20 Check valve washer - 3 off
21 Check valve gasket - 3 off

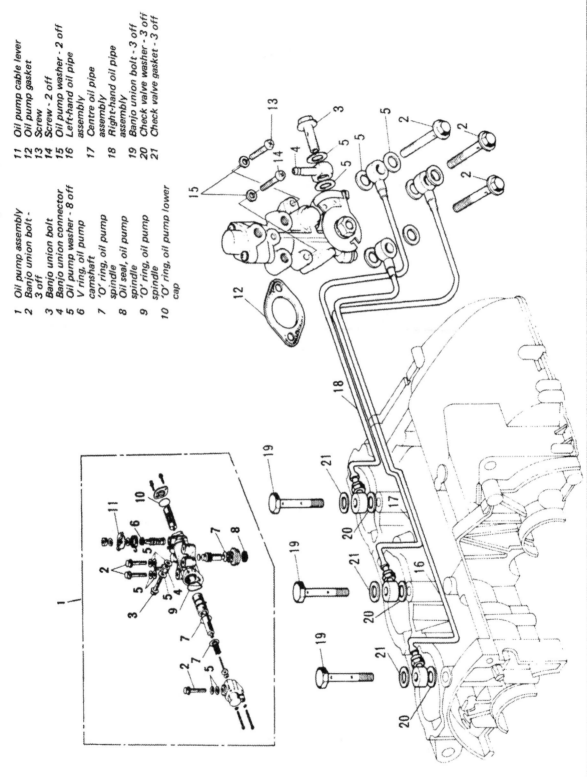

10.2 Detachable baffles are held in position with bolts

10.3 Pull out the baffle tube

11.2 Remove the side cover

11.3 Ease out the air cleaner assembly

11.4 For cleaning undo the mounting screws

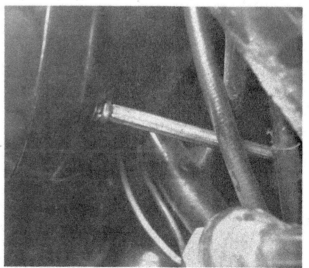

11.5 Remove the three base plate mounting screws

case sediment must be kept out of the oil pump.

5   Both fuel and engine oil caps should be examined and the air
vents cleared if clogged. Ensure the oil tank vent tube is clear and
serviceable when refitting.

6   The engine lubrication oil used is two-stroke engine oil.

---

13 Gearbox lubrication

1   The right hand cover and crankcase contains oil for lubrication
and cooling of the clutch and the various gears in the gearbox.

2   Care must be taken if refitting the right hand cover to ensure
that the gasket is new and torque tightened correctly otherwise
oil may leak.

3   The filler cap has a dipstick combined.

4   The gearbox can be drained by removing the base plug.

5   Fill with SAE 10W/30 or 10W/40 oil.

12.2 Lift off the right-hand cover

12.6 The engine lubrication oil used is the two stroke grade

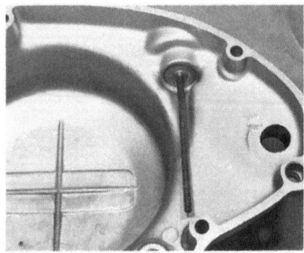

13.3 The filler cap has an integral dipstick

13.4 The gearbox oil can be drained by removing the base plug

13.5 Fill with SAE 10W 30 or 10W 40

14 Oil pump - removal, examination and reassembly

1   The oil pump can be removed and replaced with the engine in the frame. It is recommended however that the oil pump should not be removed unless it shows signs of failure. Spare parts can be obtained from a main dealer but normally the repair should be carried out by a specialist. The pump is the conventional plunger type. To replace the pump proceed as follows:

2   On the left hand side of the machine remove the tachometer cable by uncoupling the ring nut with grip pliers.

3   Remove the oil pump cover.

4   Remove the oil pump cable holder by unscrewing the two crosshead screws securing it to the case.

5   Lift the cable end off the lever and remove the cable. Pull the cable out from the pump area.

6   Remove the oil inlet pipe at the oil pump (black tube). Squeeze the black pipe when removing it to prevent loss of oil then take the end of the pipe well above the oil tank and secure. The end can be sealed if necessary whilst the pump is changed.

7   The control lever can be removed for replacement if damaged by removing the nut securing it to the pump but normally keep it with the pump. Remove the crosshead screws securing the pump to the cover and remove the pump.

8   Remove the three outlet pipes.

9   To refit reverse the procedure.

15 Adjusting the oil pump

1   Fully close the throttle.

2   Turn the oil pump adjuster so that the marks on the lever stop and the pump control lever align.

3   The illustration shows the cable before alignment. With the throttle closed use the oil pump cable adjuster to align the marks.

4   Run the engine at low speed and check for any leakage of the unions at the oil pump and also at the three crankcase connections. Our illustration shows these connections revealed during the stripdown of the engine.

14.2 Remove the tachometer cable

14.3 Remove the oil pump cover

14.4 Remove the oil pump cable holder

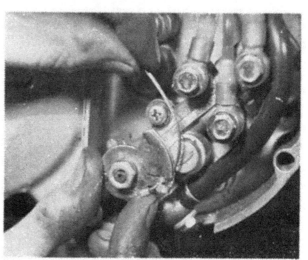

14.5a Lift the cable end off the lever

14.5b Remove the cable

14.5c Remove the cable from the pump area

14.6 Remove the oil inlet tube

14.7 Remove the control lever

14.8 Remove the three outlet pipes

15.2 The marks on the lever stop and the pump control lever must coincide

16 Checking the oil feed

1 After an engine has been rebuilt, or the oil pump disturbed, it is essential to check the oil feed. If there is no oil pressure the engine will be starved of oil leading to a seizure. Air during disassembly can become trapped inside the pipes and impede the flow of lubricant.

2 Ensure oil is available from the pump inlet pipe before connecting. Ensure this is a flow not just oil remaining in the pipe. The feed pipe must be fully primed.

3 Idle the machine just below 2000 rpm and hold the oil pump throttle control lever fully open to remove the air bubbles from the three plastic outlet feed pipes. Loosen the banjo bolt to permit the escape of air. You will lose a small amount of oil in the process so have a cloth under the pipes to catch the drips.

4 If oil bubbles persist check all the connections in the engine oil lubrication system as far as the banjo bolt connections on the crankcase. The left side oil banjo bolt to the crankcase is accessible when the machine is fully assembled. Air bubbles could also be detected at this point.

16.3 The three crankcase connections. Remove the air bubbles from the three plastic feed pipes

16.4 As far as the banjo bolt connections in the crankcase

17 Fault diagnosis

Symptom	Cause	Remedy
Engine slowly fades and stops	Fuel starvation	Sediment in filter bowl or blocking float needle - clean fuel system. Blocked vent hole in filler cap - clear. Leaking vacuum pipe from carburettor to petrol tap - replace pipe. Leak at carburettor bowl gasket.
Engine runs badly, black smoke from exhausts	Flooding carburettors	Dismantle and clean carburettors. Look for punctured float or leaking needle valve.
Engine lacks response and overheats. Acceleration below standard with loss of power	Weak mixture. Air cleaner disconnected or hoses split	Check for partial blockage in fuel/carburettors. Adjust carburettors. Repair.
Exhaust emits white smoke	Too much oil. Incorrect grade of oil	Reset oil pump. Drain completely and refill.
Lack of response to throttle variations	Exhaust fouled up	Remove and service silencer baffles.
Starts badly when hot and/or performance falls of when hot	Mixture too rich	Adjust carburettors
Engine difficult to start and runs badly	Crankcase air leaks	Strip engine and renew defective oil seals.

Chapter 3 Ignition system

Contents

Specifications

Spark plugs

Make	NGK
Type	B-9HCS
Alternative	Champion L77 J

Plug gap:

'S' series	0.016 - 0.020 in (0.40 - 0.50 mm)
Installation torque	18.1 - 21.7 ft lbs (2.5 - 3.0 kg m)

Contact breakers

Gap	0.012 - 0.016 in (0.30 - 0.40 mm)

Ignition timing	23° BTDC
Capacitor	0.18 - 0.25 microfarad
Resistance	5 meg ohms

1 General description

The spark necessary to ignite the petrol/air mixture in the combustion chambers is derived from the 12 volt battery and three ignition coils. Each of the three cylinders operates with its individual contact breaker, capacitor and coil to determine the exact moment at which the spark will occur. When the contact breaker points separate the 12 volt low tension circuit is broken and a high tension voltage peaking at about 15,000 volts is developed across the coil passing across the 0.016 inch spark plug air gap to earth. This spark will ignite the compressed petrol/air mixture in the combustion chamber. In the 'S' series machines the system operates with each plug being fired at rotation intervals of 120°. This gives three firings per rotation of the crankshaft.

The ignition circuit is operated by an ignition switch in the centre of the steering head between the speedometer and tachometer heads. It is activated by an ignition key.

The three contact breakers and their associated capacitors form part of the stator plate assembly surrounding the alternator rotor.

Instructions on the removal of the alternator are found in Chapter 1, Section 10.

When servicing the ignition and electrical sections a mini multimeter is extremely useful and could well pay for itself in very short time.

2 Crankshaft alternator - checking the output

Checking the output of the alternator is a job for a specialist unless you possess some electrical test equipment. It is advisable to carry out the simple checks outlined in Chapter 6.4 if you think they are within your technical capability.

3 Ignition coils - checking

The ignition coils are sealed units and need little attention apart from ensuring that they are mounted cleanly and tightly and that the connections are sound. If you suspect a coil and have a meter you can make a continuity check comparing a known good coil against the suspect one. It is highly unlikely that all three coils will fail at once so in that case look for the fault elsewhere. The three coils are removed by uncoupling their individual clips after disconnecting the electrical connections. They are located close to the front forks.

4 Contact breakers - adjustment

1 Remove the contact breaker cover on the left hand side of the machine by undoing the securing screws to gain access to the three

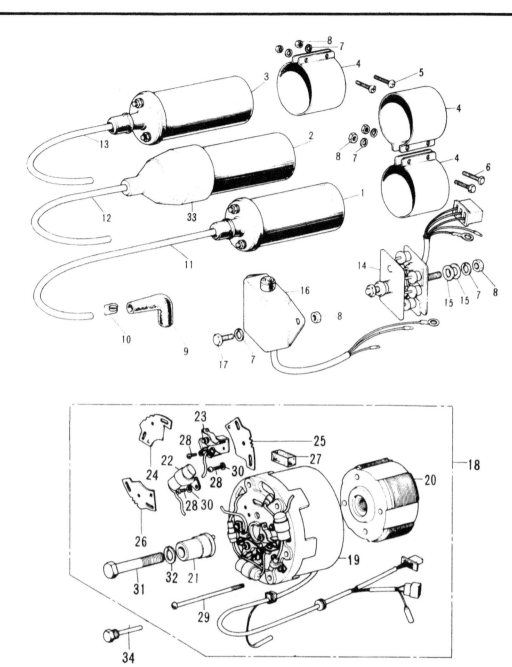

Fig. 3.1. Alternator and ignition system

1 Ignition coil, right-hand
2 Ignition coil, centre
3 Ignition coil, left-hand
4 Ignition coil band - 3 off
5 Screw - 2 off
6 Bolt - 2 off
7 Spring washer - 7 off
8 Nut - 7 off
9 Spark plug cap - 3 off

10 Spark plug cap spring - 3 off
11 High tension lead (plug lead), left-hand
12 High tension lead (plug lead), centre
13 High tension lead (plug lead), right-hand
14 Rectifier
15 Plain washer - 4 off
16 Voltage regulator
17 Bolt

18 Alternator complete
19 Stator assembly
20 Rotor
21 Contact breaker cam
22 Condenser - 3 off
23 Contact breaker points - 3 off
24 Contact breaker plate, left-hand
25 Contact breaker plate, centre
26 Contact breaker plate,

right-hand
27 Lubricating felt - 2 off
28 Screw - 12 off
29 Screw - 3 off
30 Spring washer - 12 off
31 Bolt
32 Spring washer
33 Ignition coil cap - 3 off
34 Rotor puller (service tool)

sets of breaker points. They are marked L.C.R.

2 Turn the engine over slowly using the kickstarter until one set of points is open to its widest point.

3 Examine the contact faces and if they are pitted or burned they will require further attention.

4 Measure the points gap using a feeler gauge. If the gap is 0.012 to 0.016 inch they are within specification.

5 If the gap is not correct loosen the point mounting screw slightly. Insert a screwdriver into the adjacent slot and using a feeler gauge blade of 0.014 thickness adjust the gap by manipulating the screwdriver. Tighten the mounting screw and recheck the gap. The feeler gauge blade must be a good sliding fit.

6 Repeat this adjustment for the other two sets of breaker points.

7 Before replacing the cover, place a very light smear of grease on the contact breaker cam keeping well clear of the points.

8 Apply a small amount of oil to the felt pads.

1.3 The three contact breakers and their associated capacitors form part of the stator plate assembly

5 Contact breaker points - removal, renovation and replacement

1 If the contact breakers are burned, pitted or badly worn, they should be removed for attention. If they are in a bad condition they should be renewed. They are manufactured from high quality tungsten steel which has the twin virtues of being durable and electrically excellent.

2 The points should be dressed by giving them a few strokes of a points file. When using this special file it is essential not to remove a lot of metal and keep the points absolutely square. Remove any dust with a suitable solvent and finish by drawing a piece of white paper through the gap several times until you are satisfied they are clean. Rotate the engine and recheck.

3 Do not forget you have three sets of points to check.

4 To remove the points, first undo the mounting screw then disconnect the wire to the condenser. Replacement is the reverse; always re-adjust after refitting.

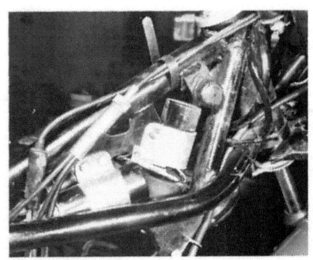

3.1 The coils in their clips

6 Capacitors - removal and replacement

1 Three capacitors are included in the contact breaker circuitry to prevent arcing across the contact breaker points on separation. Each capacitor is connected in parallel with its own set of points and if a fault develops, ignition failure will occur in that particular circuit.

2 If the engine is difficult to start or if misfiring occurs on one cylinder, it is possible that the capacitor in the ignition circuit of that cylinder has failed. A quick check is to separate the contact breaker points by hand with the ignition on. If a spark occurs across the points and they have a blackened and burnt appearance, the capacitor can be regarded as unserviceable. Arcing between the contact breaker points, points to a capacitor failure or bad and dirty connections in that area.

3 A suspect capacitor is a 'throw-away' item. Replacement is the safer bet. Auto-electric experts can use a capacity checker to read 0.18 - 0.25 microfarad or an insulation resistance of greater than 5 megohms. A capacitor can be charged from the 12 volt battery then shorted to see if a nice fat spark can be produced for a serviceable item. Do not create sparks in a confined space with petrol vapour around. The condenser must be electrically isolated during these tests.

4 To replace a capacitor remove the screw passing through an integral clamp to the contact breaker baseplate. It is also necessary to remove the capacitor lead wire attachment to the terminal or the moving contact breaker arm.

5 It is highly improbable that three capacitors will fail at the same time so look for the cause elsewhere if it seems at first that this is the trouble.

6 When replacing check that the insulating washers at the capacitor lead terminal connection are in the correct order to prevent electrical isolation and the recurrence of arcing across the points.

4.1 Remove the breaker point cover

4.4 Measure the points gap using a feeler gauge

4.5 Loosen the mounting screw and adjust gap

7 Ignition timing - checking and resetting

1   Good performance is dependent upon the accuracy with which the ignition timing is set. Even a small error can cause a marked reduction in performance and damage to the pistons. Although alignment of the cylinder/stator timing marks will verify whether the timing is accurate within certain limits it is possible to have it cross checked with a dial gauge and a lamp or meter. The lamp device can be fabricated from a torch battery, bulb, two small lengths of light wire and a couple of crocodile clips. It is assumed however for this exercise that a meter is connected across the points to read voltage when they open. The timing is set individually for each cylinder.
2   Connect the meter across the selected set of points to read DC volts up to 15 volts.
3   Switch on the ignition to supply voltage to the points.
4   Locate the viewing aperture to the cylinder 'F' marks. There will be three of these - marked RF (right cylinder F mark), LF (left cylinder F mark), CF (centre cylinder F mark).
5   Turn the engine over until the line stamped on the stator assembly lines up with the appropriate 'F' mark for the points being voltage checked.
6   Loosen the timing plate screws and adjust the timing plate by screwdriver in the timing slots until the voltmeter shows a voltage.
7   Retighten the plate screws and carry out the exercise on the other two cylinders.
8   Turn the engine and check that as the timing lines align the appropriate set of points opens.
9   Remember you have three sets of timing plate screws, three plug point adjustments, three 'F' mark alignments and three sets of contacts to attend to!

8 Spark plugs - checking and resetting the gaps

1   Each of the triple cylinder machines is fitted with three NGK spark plugs. The standard plug to be fitted is an NGK B-9HCS with an alternative Champion L77J. There is a facility for spare plugs to be carried in the stowage under the seat. The gaps should be set at 0.016 to 0.020 inch (0.40 to 0.50 mm).

7.3 Locate viewing aperture to the cylinder 'F' marks

7.4 Lines up with the appropriate 'F' mark being checked

Excessive black deposits caused by over-rich mixture or wrong heat value

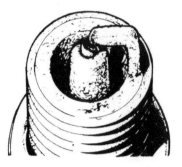

Mild white deposits and electrode burnt indicating too weak a fuel mixture

Plug in sound condition with light greyish brown deposits

Fig. 3.2a Spark plug maintenance

White deposits and damaged porcelain insulation indicating overheating

Broken porcelain insulation due to bent central electrode

Electrodes burnt away due to wrong heat value or chronic pre-ignition (pinking)

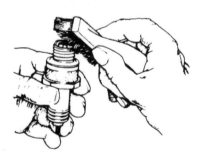

Cleaning deposits from electrodes and surrounding area using a fine wire brush

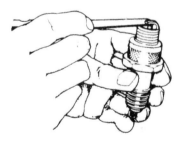

Checking plug gap with feeler gauges

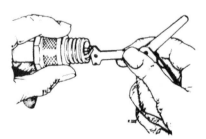

Altering the plug gap. Note use of correct tool

Fig. 3.2b Spark plug electrode conditions

7.5 Loosen the timing plate screws and adjust the timing plate by screwdriver

2 Operating conditions may dictate a change in spark plug grade. Mild conditions may call for plugs one step hotter. Severe riding conditions may require one step colder. Always use the grade of plug recommended by the manufacturer of the machine and if these are not available the dealer will recommend available alternatives.

3 Check and clean the gaps at every monthly or 500 mile service as the spark plug works hard on a high performance two-stroke

such as the Kawasaki. Refer to Fig 3 on spark plug maintenance. Note the following points:

4 Always identify the plug from each of the three cylinders - it will help in diagnosing particular engine conditions.

5 Clean dirt and grease from around the area before unscrewing the plug as this dirt could fall into the open plug hole.

6 If the plug seems immovable try a little penetrating oil around the base and leave for a short period. Do not break the plug off with excess zeal. Use a 13/16 inch deep socket - a new machine should have one in the tool kit.

7 After plug removal clean deposits from the electrodes and surrounding area using a fine wire brush.

8 To reset the gap, bend the outer electrode to bring it closer to the central electrode and ensure that the correct feeler gauge blade can be inserted.

9 Use the correct plug gap tool such as the one illustrated in Fig 3.

10 Never bend the centre electrode or it will break.

11 When refitting, screw in with the hands then torque tighten to 18 - 22 ft lbs. Do not overtighten.

12 Dirty plugs can be sandblasted but it is often cheaper to replace. Constant sandblasting does increase the wear on the insulator.

13 The central electrode if rounded at the top can be filed flat with a plug file. The side electrode can have any round edges taken off. After filing clean away all dust using cleaning solvent.

14 If the threads in the cylinder head are stripped by the use of excessive force it is possible for your dealer to fit a Helicoil insert. If in doubt have this fitted rather than risk a plug blow-out.

15 Make sure the plug insulating caps are a good fit and have their rubber seals. These caps hold the suppressors to prevent radio and TV interference.

---

9 Fault diagnosis

Symptom	Cause	Remedy
Engine will not start	Flat battery	Remove and charge. (If fault soon returns suspect battery or charging system)
	Wiring short or break	Investigate wiring.
	Plugs oiled up	Service.
	Points fouled	Service.
	Fuse blown	Replace after rectifying fault.
Engine misfires and then stops	Whiskered spark plugs	Use a hotter grade.
	Oiled spark plugs	Use a softer grade.
	Capacitor failure	Replace.
	Points pitted	Replace.
	Coil failure	Replace.
Engine lacks response and overheating	Reduced contact breaker gaps	Check and reset gaps.
Engine fades under heavy load	Pre-ignition	Replace plugs - use correct grades.
Engine 'pinks' on load	Ignition over-advanced	Reset ignition timing.

Chapter 4 Frame and forks

Contents

1 General description

The triple cylinder 'S' series machines are fitted with double-cradle frames. They are equipped with exceptionally strong steering heads designed to provide the degree of rigidity essential for high speed riding.

The rear suspension utilises a swinging arm fork controlled by hydraulically damped rear suspension units. The front of the swing arm has a pivot shaft attached to the frame. The rear part of the swing arm is attached to the shock absorbers and as these operate the arm revolves on the pivot shaft.

The Ceriani type front forks cushion the rider from any shocks experienced by the wheels due to rough ground and bumps. They are of the telescopic type with integral oil filled one-way damper units.

2 Front forks - removal from frame

1 Remove the front wheel (as per Chapter 5.2) and position the machine on the centre stand.
2 Remove the front mudguard by withdrawing the bolts on the inside of the left and right fork legs.
3 Disconnect the headlamp wiring at the connectors and remove the headlamp.
4 Disconnect the cables connected to the handlebar controls.
5 Remove the handlebar clamps from the upper fork yoke and lift away the handlebars. It may be necessary to slacken the steering damper knob to give sufficient clearance.
6 Remove the petrol tank.
7 Remove the ignition switch cable connector which is now accessible.
8 Remove the speedometer and tachometer cables and disconnect the meters from the mounting plate. If you disconnect the meter lamp cables remove the lamps to a safe stowage.
9 At the lower end of the steering damper remove the split pin,

using pliers.
10 Turn the damper knob counterclockwise to remove it from the assembly.
11 Use a hook spanner to loosen and remove the steering head stem nut. There is one in the tool kit.
12 Remove the upper fork yoke after removing the pinch bolts. Make provision to catch the uncaged steel balls which will be released as the steering head bearings separate.
13 Pull out the front forks and covers after removing the pinch bolts through the lower fork yoke.
14 The forks can be removed for service or replacement leaving the steering head in situ as our demonstration illustrations show. In this case items 3, 4, 5, 6, 8, 9, 10 and 11 can be omitted.
15 The forks can be drained of oil by either of two methods. Each fork has a drain screw at the extreme bottom and this can be removed.
16 The second method is to take out the top bolt and pour the oil into a container after each fork leg is separated from the yokes.
17 Slacken the pinch bolts through the upper and lower yokes and withdraw each fork leg assembly downwards.

3 Front fork - dismantling and examination

1 Shock absorption for the front wheel is provided by the fork springs within the front forks. Damping action is provided by the resistance of the flow of oil inside each lower fork leg.
2 It is advisable to dismantle each fork leg separately using an identical procedure.
3 With the fork on the bench remove the top bolt, the spacer and spring guide and spring. If not removed earlier, drain the oil.
4 Fit the fork leg in a vice at the lower end. Wrap the threaded collar with a length of inner tube and with a chain wrench loosen the collar.
5 Pull the stanchion in and out of the lower leg to displace it

complete with bushes.

6   Examine the outer surface of the stanchion for damage. This damage could cause oil loss. Examine the stanchion nut and O ring for damage and replace if necessary.

7   Examine the dust seal and replace it if it is damaged or hardened. Any dirt or dust bypassing the dust seal will damage the sliding surfaces of the tube and also the oil seal.

8   The springs compress as they wear. If you have to replace one, replace the other as well to keep the individual fork legs balanced.

9   Measure the length of the springs. On earlier models of S2 and all S1 they should be 14.21 inch (361 mm). Replace at 13.78 inch (350 mm).

10 On other models S2, S3 (1973) the springs should measure 10.18 inch (258.5 mm). Replace these at 9.76 inch (248 mm).

11 The spring holder and damper unit is not a replaceable item and if damaged a new lower fork leg will be needed. Examine the seat on the upper part of the spring holder.

12 The oil seal is housed in the screwed collar. If the seal shows signs of deterioration it should be replaced.

13 Particular attention should be paid to each fork bush. These are the parts most liable to wear over an extended period of service. Worn fork bushes will cause judder when the front brake is applied and the increased amount of play can be detected by pulling and pushing on the handlebars when the front brake is on full. Replace them if in the slightest doubt.

14 The lower bush is permanently attached to the bottom end of the fork stanchion by means of a peg. The damper assembly is also attached to the lower fork leg and if it malfunctions, the fork leg should be renewed. It cannot be renewed as a separate item. The replacement of the lower bush is a job for a Kawasaki dealer.

15 Reassemble the fork leg components into the lower fork leg, in reverse order of disassembly.

4 Filling the front forks

1   It is essential that the correct amount of oil is injected into the front forks.

2.2 Remove the front mudguard

2.5 Slacken and remove the handlebar clamps

2.14 Removing the forks with the steering head in situ

2.15 Each fork has a drain screw at the base

2.17a Slacken the bolts in the lower fork yoke

2.17b Also the bolts in the upper yoke

3.3 Remove the spacer

3.3a Remove the spring guide and spring

3.4 With a chain wrench loosen the two tubes

3.5 Pull the stanchion in and out of the fork leg to displace it complete with bushes

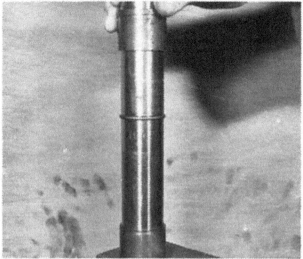

3.6 Examine the outer surface of the stanchion and the bushes

3.6a Examine the tube nut for damage

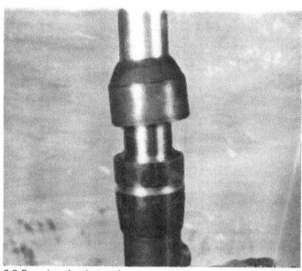

3.7 Examine the dust seal

3.9 Measure the springs

3.11 The internal damper unit

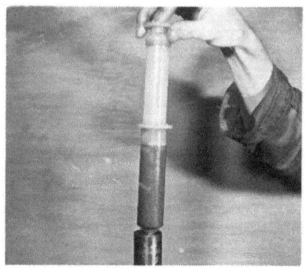

4.1 Refill the fork leg with fresh oil

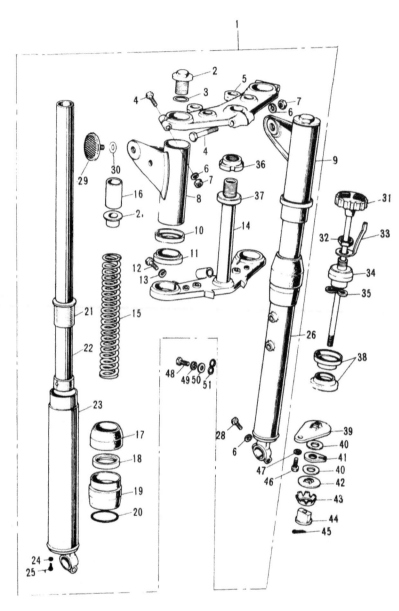

Fig. 4.1. Front forks

1 Front fork assembly
  complete
2 Fork top bolt - 2 off
3 Fork top 'O' ring -
  2 off
4 Bolt - 3 off
5 Upper fork yoke
6 Spring washer - 4 off
7 Nut - 3 off
8 Left-hand fork shroud
9 Right-hand fork shroud
10 Fork shroud washer - 2 off
11 Fork shroud gasket - 2 off
12 Bolt - 2 off
13 Spring washer - 2 off
14 Steering head stem
15 Fork spring - 2 off
16 Fork spring holder - 2 off
17 Dust cap - 2 off
18 Oil seal - 2 off
19 Lower fork leg collar -
   2 off
20 'O' ring seal - 2 off
21 Upper fork bush - 2 off
22 Stanchion - 2 off
23 Left-hand lower fork leg
24 Drain plug washer - 2 off
25 Screw - 2 off
26 Right-hand lower fork leg
27 Fork spring seating - 2 off
28 Clamp bolt
29 Reflector - 2 off
30 Reflector rubber - 2 off
31 Steering damper knob
32 Nut
33 Steering damper spring
34 Steering head stem nut
35 Crinkle washer
36 Steering head stem locknut
37 Steering head stem dust
   cap
38 Steering head bearing
   cone - 2 off
39 Steering damper anchor plate
40 Steering damper friction
   plate - 2 off
41 Steering damper stop
   plate
42 Steering damper plain
   plate
43 Steering damper spring
44 Steering damper guide
   nut
45 Split pin
46 Bolt
47 Spring washer
48 Bolt
49 Spring washer
50 Plain washer
51 Cable clamp

2   The correct oil is SAE 10.
3   If disassembling a fork, drain the existing oil.
4   On the demonstration machine 210 cc of oil was checked as
being in the fork and 210 cc of new oil was injected using a large
domestic syringe. The fork in this case had not been previously
overhauled.
5   If the fork appears noisy lift the front wheel off the ground.
6   Unscrew the top bolt and check oil level from the top of the
fork.
7   Quantities of oil for each fork leg are as follows:

S1 - S2 — 210 cc (7.1 oz)
S2 - S3 (1973 onwards) — 155 cc (5.24 oz)

8   Fluid level from top of fork with fork off the ground:

S1 - S2 — 14.75 inch (375 mm)
S2 - S3 (1973 onwards) — 13.98 inch (355 mm)

9   The front fork oil should be replaced at least every 6000
miles (10,000 km). Drain from the bottom of the fork to obviate
removal of the whole unit.

5 Steering head stem and races - examination and reassembly

1   Before commencing to reassemble the forks examine the
steering head races.
2   Examine both inner and outer races for unusual wear or
damage. The balls are a ¼ inch and there are 19 in each bearing.
3   Check the balls for wear or cracks.
4   Renew them if at all dissatisfied with either balls or race.
5   The steering head stem is removed from the frame pipe by
using a hook spanner.
6   Great care is needed if you wish to tap out the races. Use a
light hammer and a drift and tap evenly all round.
7   Examine the steering stem for damage or linearity.
8   To reassemble the stem, first fit the upper and lower cones
and the lower cup taking great care to exert even pressure around
each circumference.
9   Put a liberal amount of grease in the bearings, now in the
steering head, and insert the balls.
10 Push the steering head stem up through the bottom of the
steering head and fit the upper cone over the top of the steering
head stem.
11 Fit the locknut and use great care when tightening. There
should be smooth steering movement without play.

6 Handlebars - removal and examination

1   The handlebars are made of drawn steel tubing. It is unlikely
that you will need to renew them unless they are crash damaged.
2   If you are attracted by an alternative shape the following is
the procedure for replacing the existing pattern.
3   Remove the various cables attached to the handlebar controls.
These are clutch cable, front brake cable, starter cable and throttle
cable.
4   Disconnect the electrical connections at the cable ends and
leave the cables attached to the bar.
5   The handlebars are held by two mountings, one each side of
the damper knob.
6   Remove the bolts and lift off carefully, ensuring that no
attachment has been omitted.
7   Replace the mountings and bolts back into the lower clamps,
which are rubber mounted, onto the upper fork yoke. Examine
the rubber to ensure there are no cracks or perishing.
8   If you are replacing handlebars you now have the task of
transferring the handlebar components, which you have retained
on the old bar, to the new bar.
9   If you are buying a new type of bar, ensure first of all that
your existing controls will fit, and that the electrical and
mechanical cables are long enough for the new handlebar.

10 Do not omit to re-adjust the clutch, throttle, start and front
brake cables and ensure that the horn, turn signal and headlamp
switches are operating. These are covered in the appropriate
Sections of this manual.
11 If you are refitting the original handlebars examine them
carefully for cracks or bends.
12 Assemble in the reverse order of disassembly.
13 Remember when refitting the handlebars into the mounts to
set the angle correctly for a comfortable riding position.
14 To remove the throttle and starter cables, dismantle the twist
grip. This is in two parts, a top and a bottom. The two cables lift
out of the bottom half.
15 When reassembling the two sections of the twist grip the
return action can be adjusted by means of a screw located on the
underside of the unit. This will prevent the grip from closing of
its own accord when hand pressure is released.

7 Front forks - replacement

1   Ensure the oil seal and O ring in the outer tube have been
replaced.
2   Loosen the steering head stem nut.
3   Fit fork covers and insert each fork up through the stem
until it is even with the upper surface of the stem holding it in
place with the steering stem bolt. Tighten the steering stem nut.
4   Ensure the tops of the fork tubes are aligned with the upper
surface of the stem head.
5   Tighten lower stem bolts.
6   Tighten the steering head bolts.
7   Move the forks backward and forward in the stem to ensure
no play. If play is found see if the steering head stem locknut
requires retightening. The reverse is applicable. If too stiff steering is found loosen the locknut until a suitable movement is
evident.
8   Lift the front wheel off the ground and manipulate the handlebar to ensure easy freedom of movement.

8 Steering head lock

1   The steering head lock is secured to the lower yoke of the
forks. When in a locked position a tongue extends from the body
of the lock when the handlebars are on full lock in either direction. It locates in the notch of a plate welded to the base of the
steering head. It is impossible to ride the machine in this condition.

8.1 The steering head lock locates in a plate welded to the
steering head

2 Keep the lock lubricated as it cannot be repaired. The key on new machines will operate both the ignition switch and steering head lock. If you change either you will have to use a key for each unless you can replace with an identical mechanism.

9 Steering damper - general

1 The 'S' series machines have a steering damper fitted to provide means of adding friction to the steering head assembly so that the front forks will turn less easily. The damper can be adjusted to suit the requirements of the rider, giving in effect a fine control of movement.
2 An anchor plate is linked to the frame by a projection fitted into the anchor plate notch. Friction is provided by steel friction plates and a spring in the base. When the damper knob is tightened the spring forces the friction plates and anchor plate harder together to stiffen the steering.
3 Under normal conditions the damper should be fairly slack but for high speed or bumpy surface riding there is a need for more damper friction.
4 It is unlikely that the steering damper will require attention.
5 To remove the friction assembly it is necessary to remove the split pin securing the baseplate.
6 Turn the damper knob counterclockwise. The threaded rod operated by the damper knob will unscrew from the guide nut releasing the friction plates, spring etc.
7 When refitting the damping mechanism ensure that the projection on the frame and the notch in the anchor plate are mated.

9.6 The damper friction assembly showing split pin, spring and anchor plate projection

10 Frame - examination and renovation

1 The frame is unlikely to require attention unless accident damage has occurred. In some cases replacement of the frame is the only satisfactory course of action if it is badly out of alignment.
2 Only a few frame repair specialists have the jigs and mandrels necessary for resetting the frame to the required standard of accuracy and even then there is no easy means of assessing to what extent the frame may have been overstressed.
3 After the machine has covered a good mileage it is advisable to examine the frame closely for signs of cracking or splitting at the welded joints. Rust can also be a danger at these joints.
4 Minor damage can be repaired by brazing or welding.
5 If a frame is out of alignment it will create handling problems and if misalignment is suspect after a mishap it will be necessary to completely strip the machine for an alignment check.

11 Swinging arm rear fork - dismantling, examination and renovation

1 Remove the rear wheel, coupling and left silencer as described in Chapter 5.2 of this manual.
2 Remove the chainguard.
3 Remove the bolts at the base of the shock absorbers.
4 Remove the self locknut on the pivot shaft and slide out the shaft itself.
5 Remove the swinging arm.
6 On the bench remove the end caps and O rings.
7 Carefully extract the short sleeves.
8 Remove the long sleeve.
9 Do not remove the bushes unless out of specification or damaged. If you remove any at all they must be replaced with new bushes.
10 Examine the swinging arm for cracks or damage. Check carefully for any warping or bending or welding disintegration.
11 The sleeve and bushes should also be carefully checked for wear and damage. Measure the inner dimension of the bushes and the outer dimension of the sleeves.

Dimensions:

Item	Model	Normal	Replace at
Sleeve outer diameter	S1	0.8653 - 0.8661 in (21.979 - 22.0 mm)	0.8641 in (21.95 mm)
	S2/S3		
Bush inner diameter	S1	0.8712 - 0.8719 in (22.128 - 22.171 mm)	0.8807 in (22.37 mm)
	S2/S3	0.8673 - 0.8686 in (22.030 - 22.063 mm)	0.8780 in (22.30 mm)
Sleeve/bush clearance	S1	0.0051 - 0.0076 in (0.128 - 0.192 mm)	0.0165 in (0.42 mm)
	S2/S3	0.0012 - 0.0033 in (0.030 - 0.084 mm)	0.0139 in (0.35 mm)

12 Grease the sleeves well before refitting as damage can occur due to lack of lubrication.
13 Reassemble in the reverse order to disassembly.
14 The pivot shaft locknut should be torque tightened to 78.1 - 108 ft lbs (10.8 - 15 kg m).

11.2 Remove the chainguard

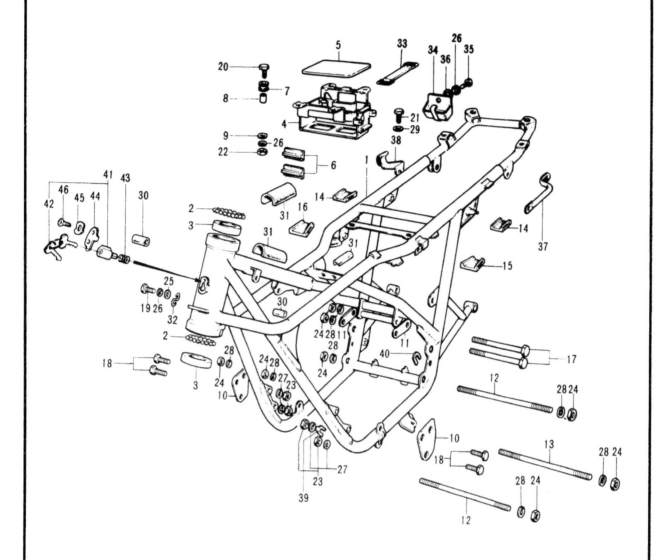

Fig. 4.2. Frame

1	Frame	10	Front engine plate - 2 off	23	Nut - 4 off	36 Plain washer
2	Ball bearing ¼ in. diam. - 38 off	11	Rear engine plate - 2 off	24	Nut - 4 off	37 Frame grip
3	Steering head bearing cup - 2 off	12	Engine mounting stud - 2 off	25	Plain washer - 3 off	38 Helmet holder bracket
4	Battery case	13	Engine mounting stud	26	Spring washer - 7 off	39 Engine fitting shim - as required
5	Battery mat	14	Damper rubber - 2 off	27	Spring washer - 4 off	40 Engine fitting shim - as required
6	Rectifier mat - 2 off	15	Damper rubber	28	Spring washer - 8 off	41 Steering head lock
7	Damper rubber - 4 off	16	Damper rubber	29	Lock washer - 2 off	42 Key for lock
8	Battery case collar - 4 off	17	Bolt - 2 off	30	Fuel tank damper rubber - 2 off	43 Steering head lock spring
9	Speedometer mounting washer - 3 off	18	Bolt - 4 off	31	Fuel tank mat - 3 off	44 Steering head lock cap
		19	Bolt	32	Cable clamp	45 Crinkle washer
		20	Bolt - 4 off	33	Tool strap	46 Self-tapping screw
		21	Screw	34	Spark plug holder	
		22	Nut - 2 off	35	Bolt	

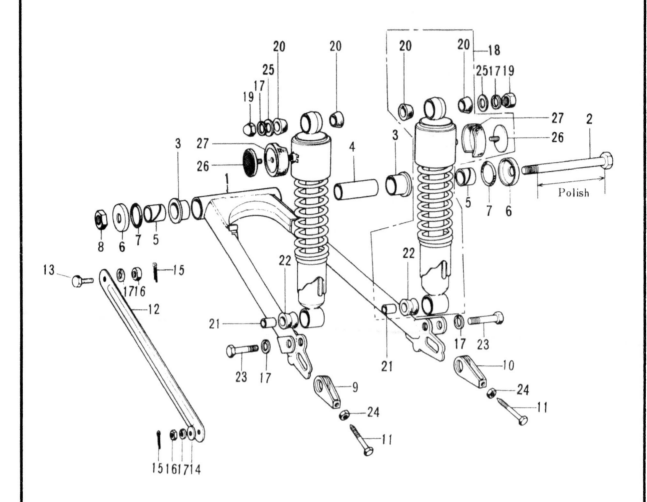

Fig. 4.3. Swinging arm and rear suspension units

1 Swinging arm
2 Swinging arm pivot spindle
3 Swinging arm bush - 2 off
4 Swinging arm sleeve
5 Swinging arm sleeve - 2 off
6 Swinging arm dust cover - 2 off
7 'O' ring seal - 2 off

8 Locknut
9 Left-hand chain adjuster
10 Right-hand chain adjuster
11 Chain adjuster bolt - 2 off
12 Rear brake torque arm
13 Torque arm anchor bolt
14 Plain washer - 2 off
15 Split pin - 2 off

16 Nut - 2 off
17 Spring washer - 6 off
18 Rear suspension unit - 2 off
19 Acorn nut - 2 off
20 Rear suspension unit rubber bush - 4 off
21 Rear suspension unit metal bush - 2 off

22 Rear suspension unit rubber bush - 2 off
23 Bolt - 2 off
24 Plain washer - 2 off
25 Plain washer - 2 off
26 Reflector - 2 off
27 Reflector rubber mount - 2 off

11.3 Remove the bolts at the base of the shock absorbers

11.4 Remove the self lock nut on the pivot shaft

11.5 Remove the swinging arm

11.6 Remove the end caps and 'O' rings

11.7 Carefully extract the short sleeves

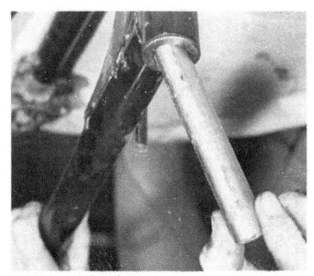

11.8 Remove the long sleeve

11.9 Renew the bushes if they are removed

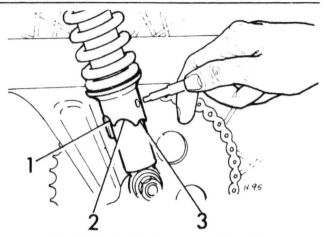

Fig. 4.4. Settings for the rear suspension units

1 Normal riding
2 High speed - solo
3 With pillion riding

12 Rear shock absorbers - examination and adjustment

1   The shock absorption system utilises springs and hydraulic dampers.
2   There are no means of draining the units or topping up because the dampers are built as sealed units. If the damping fails or if the units commence to leak the complete damper assembly must be renewed.
3   The shock absorbers can be adjusted for high speed or heavy load operation.
4   In the interests of good roadholding it is essential that both units are set to the same load setting and if you have a replacement fitted both units must be matched up.
5   The shock absorbers can be removed by removing mounting bolts at the top and base of the assembly. Protect the rubber bushes.
6   Adjustment is made by the use of a special tool or screwdriver into peg holes at the base of the spring. This lowers or raises the spring seating.
7   Clockwise rotation increases the tension and vice versa. Most tension is needed for very high speeds, a pillion passenger or when a heavy load is carried at the rear.

13 Centre stand - examination

1   The centre stand is attached to lugs welded to the bottom frame tubes and pivots on a shaft through each lug hole. A split pin secures the shaft. An extension spring is used to keep the stand in the fully retracted position when the machine is in use.
2   Check that the return spring is in good condition and that both nuts and bolts at the pivots are tight. Replace the foot rubber if damaged. Refit new split pins.
3   There is no need to stress the danger to the rider if due to faulty maintenance the stand falls whilst the machine is in motion.

14 Prop stand - examination

1   A prop stand is for use when it is not desired to use the centre stand.
2   The stand pivots from a metal plate attached to a lug on the lower left hand frame, by a bolt. It is also fitted with an extension spring to ensure the stand retracts immediately the weight of the machine is removed.
3   Ensure that the two bolts retaining the metal plate are both fully tightened and also the single pivot bolt through the eye of the stand arm. Ensure also that the extension spring is in good condition and not overstretched.
4   Replace the foot rubber if damaged.

15 Footrests - examination and renovation

1   The front and rear footrests fit onto footrest bars which bolt to gussets onto the frame at either side of the machine.
2   The rear footrests also serve as silencer mountings.
3   The footrests pivot upwards and are spring-loaded to keep them in their normal horizontal position. If an obstacle is struck, they will fold upwards thus obviating the risk of injury to the rider's foot or frame damage.
4   Damaged footrests can be disassembled. No attempt should be made to restraighten on the machine.
5   Heat can be applied when attempting to straighten the metal section but unless you have had instruction on metal 'bashing' do not attempt this. Leave it for a professional or replace the item.
6   New foot rubbers can normally be obtained from a dealer.
7   Care is necessary when replacing the front left footrest, which is adjacent to the gearchange pedal. See that the gear pedal links are at 90° angles to give adequate room for foot movement. Our illustration shows the pedal in the process of adjustment.

16 Rear brake pedal - examination and renovation

1   The rear brake pedal is attached to a pivot arm attached to the lower right hand frame tube located above the centre stand. A slot arm welded to the other end of the tube forms the attachment for the cable which operates the rear brake.
2   The tube is fitted with a strong return spring coil around its boss to provide the pedal with positive action.
3   To remove for examination:
4   Remove the clevis pin.
5   Disconnect the brake cable at rear brake.
6   Disconnect the brake light operating spring.
7   Ease the brake pedal assembly off its pivot maintaining hold on the return spring which should be safely stowed.
8   Examine the spring for any sign of weakness and replace it if at all suspect.
10 The brakes and brake light must be re-adjusted after this check.
11 If the brake pedal is bent or twisted it is an expert's job to repair. It is far better to replace it as a fractured brake pedal could be extremely dangerous.

17 Dualseat - removal and renovation

1   The seat is attached to two metal hinges on the left hand side of the subframe.
2   It opens from the right after a safety catch has been released.

3 S3 models are fitted with helmet holders that lock automatically when the seat is lowered. Helmets can be hung on hooks on the frame at the right side of the machine. They are automatically secured when the seat is lowered and locked.
5 To remove the seat for repair or recover detach at the hinges. Remove the split pins and hinge bars.
5 Three spare plugs are stowed beneath the seat.
6 The tail housing carries a tool kit and a rider's handbook.
7 To remove the stop/tail light the cables must be withdrawn through the tail housing behind the tool kit stowage.

18 Speedometer and tachometer heads - removal and replacement

1 Apart from defects in either the drive or drive cables, a speedometer or tachometer which becomes unserviceable is an instrument mechanic's job to repair.
2 A speedometer in correct working order is a statutory requirement in the UK. Apart from the legal requirement a close check on mileage recorded is a guide to service intervals.
3 To remove proceed as follows (this procedure may vary slightly with other models).
4 Uncouple the flexible cables with grips.
5 Remove the nuts that secure the indicator check mounting to the head assembly.
6 Remove the electrical connections and put the lamps in safe stowage.
7 Handle the meters with great care. A cardboard box with packing inside helps to protect them from damage.
8 Refit in the reverse order of removal.

19 Speedometer and tachometer drive cables - examination and maintenance

1 Detach both cables periodically to ensure their correct lubrication. Check also for any signs of damage to the outer covering which could lead to jerky operation.
2 When greasing withdraw the inner cable, remove the old grease, clean with a suitable solvent and thoroughly examine for damage.
3 Relubricate with the recommended high melting point grease but only up to six inches from where the cable enters either head. If this is not adhered to grease will work into the head and damage the instrument.
4 If either instrument fails first suspect a broken cable. Only a complete cable can be supplied.

20 Speedometer and tachometer drives - location and examination

1 The speedometer drive cable gear and pinion are part of the front wheel hub assembly and are driven internally from the wheel spindle. If lubricated as recommended it should not give trouble. If wear in the drive mechanism occurs, the pinion and its associated gear can be removed from the brake plate housing. The drive pinion which mates with the gear is secured within the brake plate by a circlip in front of the shaped driving plate which engages with slots in the front wheel hub.
2 The tachometer drive is taken from the oil pump assembly in which it has its own gear assembly. It is unlikely to give trouble. Always check the tachometer after adjustments in the oil pump area.

21 Cleaning the machine

1 Remove all surface dirt with a rag or sponge which is washed frequently in clean water. If the machine is particularly dirty some warm water and a mild detergent will get most of the heavy stuff off but wash off with clean water immediately. The machine should dry thoroughly and any smears on the tanks etc should be removed. Application of a good polish will enhance the machine's appearance.

15.2 The rear footrests also serve as silencer mountings

15.6 Care is necessary when replacing the front left footrest

16.7 Ease the brake pedal assembly off its pivot

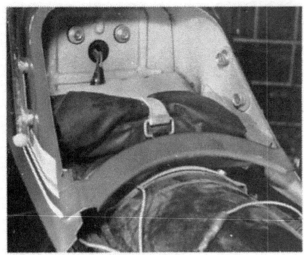

17.6 The tail housing carries a tool kit

18.4 Uncouple the flexible cable

18.6 Remove the electrical connections

2   The plated parts of the machine should be wiped clean with a damp rag. If the plated parts are corroded due perhaps to salt on roads in winter you can use one of the advertised chrome cleaners. These should have an oily base to guard against repetition of the corrosion.

3   If the engine parts are particularly oily utilise a cleansing agent such as Gunk or Jizer. Apply the agent whilst the area is dry and work it in with a brush to enable maximum penetration. Allow enough time for the agent to operate and then wash off. Do not get water into the carburettors or ignition. The polished aluminium alloy parts can now be polished with a polish to regain some of their previous sheen.

4   It is advisable to wheel the machine into a garage or shed after riding in a wet spell and dry it off. This will keep at bay the danger of rust forming. Pay particular attention to the chain which should be wiped and re-oiled to prevent water entering the rollers and causing harshness and subsequent rapid rate of wear.

22 Control cable lubrication

1   Control cables need to be regularly checked for freedom of operation and lubrication. Do not forget however to check the re-adjustment of the system when refitting the cable and do not forget that some cables have adjusters within their lengths.

2   Like other motorcycle parts control cable prices rocket and you can pay several pounds or many dollars for a replacement if you neglect the ones you have on the machine.

3   The standard method of lubrication is to remove the cables from the machine, fabricate a plasticine or plastic cover, and allow the oil to drain down the tube whilst the cable is hung from a hook. Castrol Everyman Oil is recommended for this process.

4   A Japanese manufacturer (not Kawasaki) markets in the USA a device for lubricating the cable on the machine. It consists of a rubber sealed capsule that clamps over the disconnected cable end and has an insert hole. The nozzle tube of a spray holder is then connected and the whole operation is carried out in a much shorter period.

5   A recommended cable lubricant for USA owners is Petroclean's Cable Life.

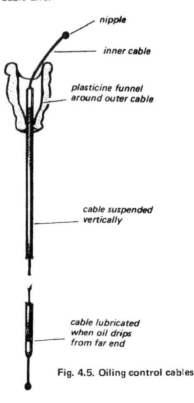

Fig. 4.5. Oiling control cables

23 Fault diagnosis

Symptom	Cause	Remedy
Machine veers to left or right with hands off handlebars	Incorrect wheel alignment Bent forks Twisted frame	Check and re-align. Check and renew. Check and renew.
Machine rolls at low speeds	Overtight steering head bearings	Slacken and reset.
Machine judders when front brake is applied	Slack steering head bearings Worn fork bushes	Tighten until play is taken up. Renew.
Machine pitches badly on uneven surfaces	Ineffective fork damping Ineffective rear shock absorbers	Check oil level. Check operation of dampers.
Fork action stiff	Fork legs out of alignment Wrong grade or amount of oil	Slacken yoke clamps, front wheel spindle and fork top bolts. Pump forks several times then tighten from bottom upwards. Drain and refill
Machine wanders. Steering imprecise. Rear wheel tends to hop	Swinging arm pivot worn	Dismantle, renew bushes and pivot shaft.

Chapter 5 Wheels, brakes and tyres

Contents

Specifications

	S1	S2	S3
Tyres			
Front (ins)	3.00 x 18	3.00 x 18	3.25 x 18
Rear (ins)	3.25 x 18	3.50 x 18	3.50 x 18
Standard type	Yokohama		
Brakes			
Front	7.1 x 1.2 ins	7.1 x 1.2 ins	10.9 ins
Type	Int. expanding	Int. expanding	Disc
Rear	7.1 x 1.2 ins	7.1 x 1.2 ins	7.1 x 1.2 ins
Type	Int. expanding	Int. expanding	Int. expanding
Tyre pressures			
Front	24 psi	24 psi	24 psi
Rear	31 psi	31 psi	31 psi
Ply rating	4	4	4

Note: Increase the rear tyre pressure by about 4 psi when a pillion passenger or very heavy load is carried. Always measure with the tyre cold. Also increase tyre pressures for sustained high speeds (over 100 mph).

Disc brake fluid	Castrol Girling Universal Brake and Clutch Fluid
	Girling — Amber
	Shell — Super Heavy Duty
	Texaco — Super Heavy Duty
	Atlas — Extra Heavy Duty
	Wagner — Lockheed Heavy Duty

Master cylinder tolerances
Cylinder inside diameter:

Standard	0.5512 - 0.5529 ins
Limit	0.5543 ins

Piston outside:

Standard	0.5495 - 0.5506 ins
Limit	0.5472 ins

Primary, secondary cup diameter:
Standard ... ... 0.5768 - 0.5965 ins
Limit ... ... 0.5709 ins
Spring length (free):
Standard ... ... 2.169 ins
Limit ... ... 1.890 ins

Recommended torque settings for disc brake components

	lb ft	kg m
Brake lever	3.75 - 5	0.5 - 0.7
Brake lever adjuster	5.9 - 8.3	0.8 - 1.2
Master cylinder clamp	4.5 - 6.25	0.63 - 0.88
Disc mounting bolts	11.6 - 16	1.6 - 2.2
Caliper mounting	19 - 23	2.5 - 3.3
Bleeder valve	5.9 - 7	0.8 - 1
Caliper shafts	22 - 26	3 - 3.6
Pressure switch	11.25 - 14.1	1.5 - 2
Brake pipe nipple	12.5 - 13	1.7 - 1.8
Banjo bolt fitting	19 - 23	2.5 - 3.3
Three-way fitting mounting	3.75 - 4.17	0.5 - 0.6

1 General description

Kawasaki 'S' models are fitted with wheels with steel rims. The wheel hubs are cast aluminium.

Late S2 and the S3 models have a 10.9 in disc brake on the front wheel.

The rest of the 'S' series machines carry 7.1 x 1.2 internal expanding brakes front and rear.

For machines with disc brakes the applicable sections in this Chapter are 14 - 22.

2 Front wheel - removal

1 This procedure applies to models without disc brakes.
2 At the front hub loosen the brake adjustment nut and remove the brake cable.
3 Take out the speedometer cable bolt and remove the speedometer cable.
4 Raise the machine on the centre stand.
5 On the other side of the wheel remove the pinch bolt in the base of the right hand lower fork leg.
6 On the left hand side of the wheel remove the split pin and then the spindle nut. Pull out the spindle and lift out the front wheel. Remove the spindle oil seal.
7 Remove the brake plate from the left hand side of the wheel. This reveals the speedometer gear and the mating pinion (fallen into the bottom of the brake plate assembly in the illustration). The brake drum is available for examination.
8 The shoes can be removed from the brake plate by lifting upwards whilst springing them apart.

3 Front brake assembly - renovation and reassembly

1 Examine the brake linings for oil, dirt or grease.
2 Surface dirt can be removed with a stiff brush but oil soaked linings must be replaced at once. High spots can be taken down with emery cloth. It is necessary to renew the brake shoes if the linings are faulty because the linings are bonded on and not supplied separately.
3 To determine the extent of wear on the brakes it is necessary to determine the thinnest part of the brake lining and measure it. This should normally be 0.20 inch (5 mm). Replace both shoes if either shoe is worn to 0.118 inch (3 mm) or less.
4 The brake return shoe springs should not be worn or stretched. Measure the length of each spring. It should be 1.85 inch (47 mm). Replace at 1.97 inch (56 mm).
5 Check the clearance between the brake camshaft and the bush in the front brake plate. It should be 0.0008 - 0.0028 inch (0.02 to 0.07 mm). Replace at 0.02 inch (0.5 mm).
6 Measure the inside diameter of the brake drum. After long

service the brake shoes (which are periodically changed) wear down by friction of the inner surface of the brake drum.
7 On S1 and S2 machines (S3 has discs) the standard diameter is 7.087 inch (180 mm). If at any point you read 7.116 inch (180.75 mm) or more you must renew the drums.
8 Reassemble the front brake in reverse order of removal.
9 The front brake operating arm must be approximately parallel to the fork leg.
10 Use grease on the brake pedal bearing and brake lever.
11 Use brake grease on the camshaft bearings in the brake panels. A white grease such as Castrol PH is recommended.

4 Wheel bearings - examination and renovation

1 The front hub which includes the brake assembly contains two bearings. These are fitted on either side of the hub.
2 The rear hub is fitted with three ball bearings.
3 On removal do not confuse the types. The front wheel bearings are 6302 and the rear wheel bearing 6205. A 'Z' after the number indicates that it is located on the brake plate side.
4 The oil seals are not interchangeable. The front wheel hub oil seals are WTC 25 42 8. The front wheel brake plate (S1, S2) oil seals are WOC 55 68 7 with the speedometer gear holder needing an OJ 32486 (S3).
5 The oil seals on the rear wheel coupling (S1, S2) are WTC 35 52 7 and PJX 135527 (S3).
6 Access is available to both of the wheel bearings when the brake plates have been removed. The oil seals can be removed for examination at the same time.
7 Remove the first bearing by hitting the bearing spacer to knock it out. The remaining bearing is removed by setting a rod to its inner side and knocking it out. Damage to the bearing seating surface can be avoided by light even tapping.
8 Place the wheel bearings in a jar with petrol to clean thoroughly.
9 After cleaning, oil the bearing before spinning to check for smooth operation.
10 The oil seals when removed with the bearings should have a close check and if in any doubt replace. Examine the main lip which prevents the leakage of hub grease and the subsidiary lip which stops dirt and water from contaminating the bearings.
11 The bearings will need packing with grease before being driven back into the hub.
12 Grease the hub and use the wooden drift to drive the bearings back into position. It is essential that the seals and bearings are at an exact right angle to the axle.

5 Front wheel - reassembly and replacement

1 The front wheel is reassembled in the reverse order to that of dismantling.

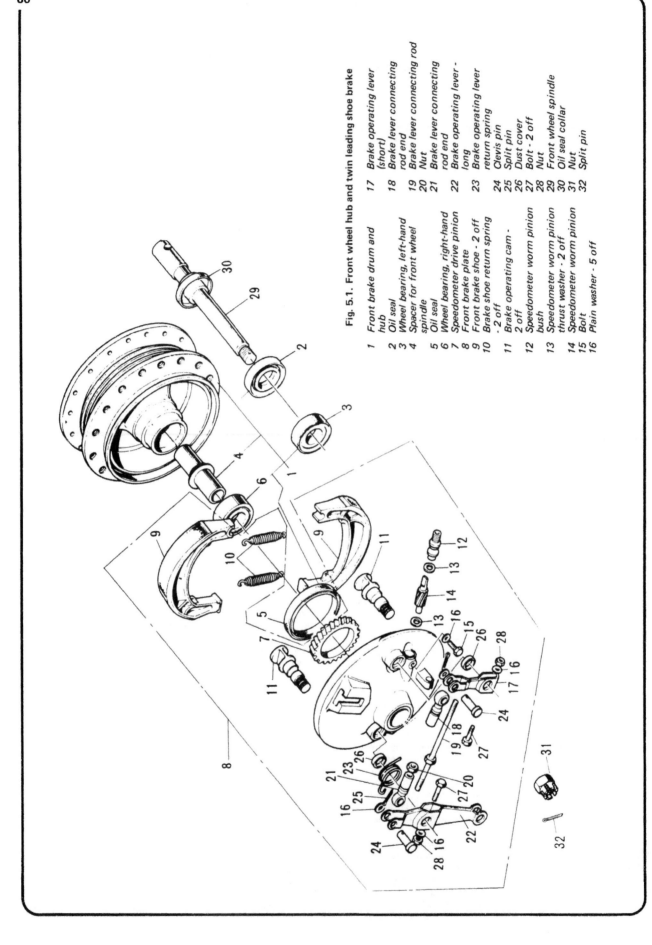

Fig. 5.1. Front wheel hub and twin leading shoe brake

1 Front brake drum and hub
2 Oil seal
3 Wheel bearing, left-hand
4 Spacer for front wheel spindle
5 Oil seal
6 Wheel bearing, right-hand
7 Speedometer drive pinion
8 Front brake plate
9 Front brake shoe - 2 off
10 Brake shoe return spring - 2 off
11 Brake operating cam - 2 off
12 Speedometer worm pinion bush
13 Speedometer worm pinion thrust washer - 2 off
14 Speedometer worm pinion
15 Bolt
16 Plain washer - 5 off
17 Brake operating lever (short)
18 Brake lever connecting rod end
19 Brake lever connecting rod end
20 Nut
21 Brake lever connecting rod end
22 Brake operating lever - long
23 Brake operating lever return spring
24 Clevis pin
25 Split pin
26 Dust cover
27 Bolt - 2 off
28 Nut
29 Front wheel spindle
30 Oil seal collar
31 Nut
32 Split pin

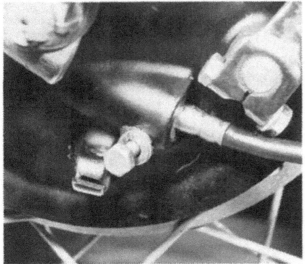

2.3 Take out the speedometer cable bolt

2.3a Remove the speedometer cable

2.5 Remove the pinch bolt through the lower fork leg

2.6a Remove the split pin and the spindle nut

2.6b Pull out the spindle

2.6c Lift out the front wheel

2.6d Remove the spindle seal

2.7 Remove the brake plate

2.8 The shoes can be removed from the brake plate

3.9 The front brake lever must be approximately parallel to the fork leg

2   Ensure that grease has been applied to the bearings, oil seal and the front panel speedometer pinion gear.

3   On no account must grease get onto the brake linings or the surface of the brake drum. If grease does get onto these parts, clean off with petrol.

4   Replace the brake plate assembly back into the drum and replace the oil seal on the opposite side of the wheel.

5   Fit the wheel back into the forks.

6   On the right side of the machine replace the spindle and then the pinch bolt.

7   On the brake assembly side fit the spindle nut and after torque tightening fit the split pin. Torque tighten the front wheel spindle nut to 48 - 61 ft lbs (6.7 - 8.5 kg m).

8   Refit the speedometer cable and then the retaining bolt for the cable. (Ensure that the cable drive has first been installed.)

9   It is now necessary to re-align and adjust the front brake.

6   Front wheel - adjusting the drum brakes

1   The front brake has twin leading shoes.

2   Each brake shoe has two operating cams.

3   The front brake stop lamp switch is fitted into the cable and is not adjustable. It is not fitted to UK models.

4   Correct adjustment is achieved if braking action commences when the front brake lever is pulled one inch (25 mm).

5   Adjustment is effected by turning the nut at the front brake cam lever.

6   The operating rod which links both operating arms of the twin leading shoes should not require adjustment unless the original setting has been disturbed or new brake shoes fitted.

7   It is imperative that the leading edge of each brake shoe contacts the brake drum simultaneously if maximum braking efficiency is to be obtained.

8   To adjust, if the original setting is lost, proceed as follows:

9   Align the first cam lever with the serrations on the camshaft creating a 90° angle to the fork leg when the brake commences to operate.

10 Fit the second cam lever parallel to the first.

11 Loosen the cam lever connecting rod and turn it one turn. If the operator is facing rearward with the spanner in the right hand to the rod he will turn the rod clockwise.

12 This leaves the front shoe free to be adjusted.

13 Lift the front wheel clear of the ground and spin the wheel.

14 With the wheel spinning tighten the adjustment nut at the front cam lever until the front brake bites very slightly.

15 Now turn the connecting rod manipulated in item 11 of this Section in the reverse direction so that the second brake shoe begins to bite.

16 Tighten the locknut.

17 It will be necessary to re-adjust the play in the front brake lever.

18 Check that the brake pulls off correctly when the handlebar lever is released. Sluggish action is normally due to a poorly lubricated control cable, broken or extended brake shoe return springs or a tendency for the brake operating cams to bind in their bushes. Binding brakes affect engine performance and can cause severe overheating of both brake shoes and the wheel bearings.

7 Front wheel - spokes and balance

1 With the wheel on the machine, place the machine on its centre stand to raise the front wheel clear of the ground.
2 Spin the wheel and check the wheel alignment by viewing the wheel in motion.
3 Small irregularities can be corrected by tightening the spokes in the affected area, although experience is necessary at this task to avoid over-correction.
4 Any flats in the wheel rim should be evident at the same time. These are more difficult to remove and in most cases it will be necessary to have the wheel rebuilt on a new rim. Apart from the effect on stability, a flat will expose the tyre bead and walls to greater risk of damage if the machine is run with a deformed wheel.
5 Examine the wheel for loose or broken spokes. Tapping the spokes is the best guide for tension. A loose spoke will produce quite a different sound from one correctly tightened.
6 Tighten a loose spoke by turning the nipple (right hand thread).
7 Always recheck for run-out by respinning the wheel.
8 Do not tighten the spokes an excessive amount. If it appears to be necessary, remove the tyre and tube so that the protruding ends of the spokes can be ground off in order to prevent them chafing the inner tube and creating punctures. The spokes should be tightened to 22 - 26 inch lbs torque (0.25 - 0.30 kg m).
9 A bent or faulty spoke can be removed by completely unscrewing the threaded portion and then renewing the bent or faulty item.
10 Balance weights can be fitted to correct wheel unbalance and unsafe riding conditions. These are added to spokes on the lighter side of the wheel. Most tyre centres will carry out a wheel balance when supplying new tyres or charge a small fee for balancing if not supplying new tyres. Balance weights can be fitted of 1/3, 2/3 and 1 oz (10, 20, 30 grams).

8 Rear wheel - removal

1 It is assumed the right hand silencers have previously been removed.
2 Disconnect the rear brake cable.
3 Loosen the chain adjustment, remove the chain link and then the chain.
4 Remove the split pin on the rear brake torque arm bolt.
5 Remove the nut and disconnect the torque arm.
6 Remove the spindle nut split pin, the spindle nut and then the spindle itself.
7 The chain adjustment collar will be free.
8 Draw out the wheel.
9 The procedure above is for the demonstration model which was a 72 - S2. It is possible that procedures will vary with later models when it will be necessary to first loosen the chain case guard mounting bolt and before removing the wheel to take out the chain adjuster stoppers and their associated bolts.
10 Remove the sleeve nut, the coupling with rear sprocket and the left hand chain adjuster.

9 Rear wheel - examination

1 The procedures for wheel balancing and spoke tightening detailed in the front wheel overhaul apply similarly to the back wheel.
2 Tyres, rear brakes and rear sprockets are also dealt with in their own sections of this Chapter.

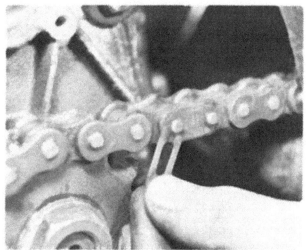

8.3 Remove the chain link

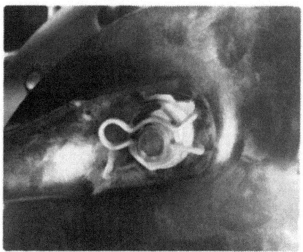

8.4 Remove the split pin on the torque arm bolt

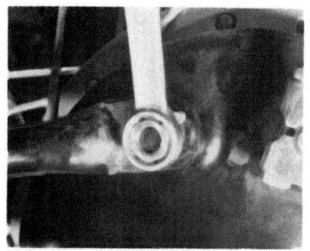

8.5 Remove the nut and disconnect the torque arm

3   With the brake assembly removed examine the inner surface
of the brake drums which after long service are worn down by
friction with the linings. Measure the diameter inside the drum at
various points. On all S models it should be 7.087 inch (180 mm)
and be replaced at 7.116 inch (180.75 mm). These are the same
dimensions as the front wheel.
4   The rear hub breaks down into the brake drum, mechanism,
plate, shoes, sprocket and coupling. The coupling is on the left
and the plate on the right of the drum.
5   Between the coupling and the rear brake drum are fitted cush
drive rubbers.
6   The cush drive rubbers can be removed for examination,
cleaning or replacement. If the rubbers are cracked or deformed
they must be replaced as they will fail to form an efficient
buffer.
7   Remove the hub bearing O ring, the ball bearing, and then the
bearing spacer.
8   The bearings and oil seals should be regreased prior to refitting but do not drop any grease on the brake linings or drum
surface. Clean off any spilt grease with petrol.

10 Rear brake assembly - examination

1   Remove the rear brake panel to gain access to the brake shoes,
cam lever etc.
2   Pull the brake shoes upwards off the pivot studs to remove
them, whilst they are sprung apart.
3   The brake linings are not replaceable by rerivetting as they are
bonded to the shoes.
4   Clean off the brake linings. Oil soaked or damaged linings
mean that you need a pair of replacement shoes. A stiff brush will
move the dirt and grit, and emery cloth the high spots.
5   Measure each brake lining at the point of greatest wear. The
thickness for rear brake linings should be 0.2 inch (5 mm) with
replacement at 0.08 inch (2 mm).
6   S3 linings are a little thicker than the S, S2 models and should
be 0.21 - 0.24 inch (5.4 - 6.1 mm) but replaced at the same 0.08
inch.
7   Check the brake springs at the same time. These should be
2.2 inch (56 mm) and replaced at 2.32 inch (59 mm).
8   The brake camshaft should be examined for 'hole' wear. If the
clearance is excessive you fail to get efficient braking action.
Measure the diameter of the camshaft against the inside diameter
of the shaft hole. These should be as follows for front and rear
camshafts:

Location	Standard	Replace at
Shaft hole diameter	0.5906 - 0.5916 in	0.6004 in
	(15 - 15.027 mm)	(15.25 mm)
Camshaft diameter	0.5899 - 0.5889 in	0.6201 in
	(14.984 - 14.957 mm)	(15.75 mm)
Clearance	0.0008 - 0.0028 in	0.0197 in
	(0.01 - 0.07 mm)	(0.50 mm)

11 Rear wheel - replacement

1   To reassemble the rear wheel back into the frame reverse the
dismantling procedure.
2   Align the wheels and adjust the chain with the chain adjusters
in the same operation, so that the rear wheel is in line with the
front wheel.
3   Tighten the rear wheel spindle to a 48 - 61 ft lb setting (6.7 -
8.5 kg m). This is the identical torque to the front wheel spindle.

12 Adjusting the rear brake

1   If the adjustment of the rear brake has been carried out correctly the brake pedal will have a travel of from 0.75 inch to 1.25
inch (20 - 30 mm). Before the amount of travel is adjusted the
brake pedal should be set so that the pedal is in the best position

8.6 And then the spindle itself

8.7 Draw out the wheel

9.6 The cush drive rubbers can be removed for examination

9.7a The hub bearing 'O' ring

9.7b Now the ball bearing

9.7c And the bearing spacer

10.1 The brake plate and brake assembly will pull off the hub

10.4 Clean off the brake linings

for instant operation. Check the position of the footrest near at hand.

2   The height of the brake pedal is determined by an adjuster at the end of the brake cable at the point where it joins the pedal arm. To raise pedal height screw inwards and vice versa.

3   The length of travel is controlled by the adjustment nut at the end of the brake lever which is at the rear of the wheel hub assembly. If the nut is screwed inwards travel is decreased and the reverse applies. When assembling the brake cam lever to the camshaft it should be mounted so that when the brakes first start to bite the brake cable will be approximately at right angles to the cam lever.

4   It could be found necessary to re-adjust the height of the stop lamp switch if the pedal height has been altered to any marked extent. The rear stop light switch should be adjusted so that the light operates when the pedal has travelled 0.625 - 0.75 inch (15 - 20 mm). Remember to adjust the nut and not the switch body which is plastic and breakable.

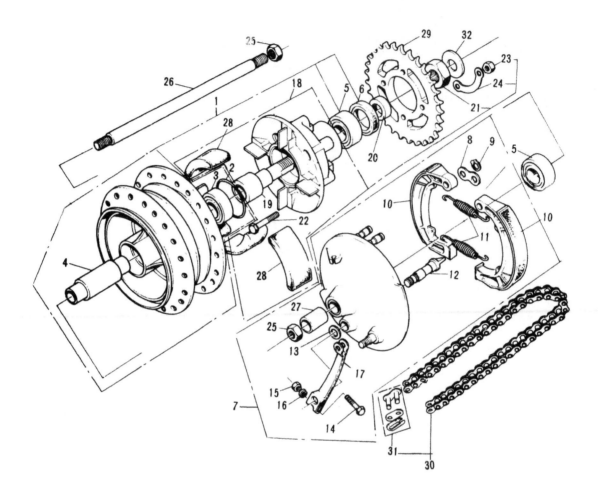

Fig. 5.2. Rear hub, brake drum and cush drive

1	Rear hub, brake drum and cush drive assembly	10	Rear brake shoe - 2 off	19	Rear wheel spindle sleeve
2	'O' ring	11	Brake shoe spring - 2 off	20	Rear spindle sleeve collar
3	Wheel bearing - 2 off	12	Rear brake operating cam	21	Sleeve nut
4	Rear hub bearing spacer	13	Brake operating cam dust shield	22	Sprocket retaining bolt - 4 off
5	Cush drive bearing	14	Bolt	23	Nut - 4 off
6	Oil seal	15	Nut	24	Sprocket lock washer - 2 off
7	Rear brake plate	16	Plain washer	25	Nut - 2 off
8	Brake shoe retaining washer	17	Rear brake operating arm	26	Rear wheel spindle
9	Circlip - 2 off	18	Cush drive assembly		

27 Right-hand wheel spindle spacer
28 Cush drive rubber - 4 off
29 Rear wheel sprocket (48 teeth, standard)
30 Final drive chain (102 rollers)
31 Spring link
32 Sleeve washer

12.2 The adjuster is at the end of the brake cable

13 Brake lining wear indicators

1   On both brake plates of the 1974 S1 models and on the rear brake plate of the S3 a brake lining wear indicator has been fitted.
2   The indicator should be regularly examined. If it shows that you have passed the usable range into the red area then the brake shoes must be replaced immediately. The other brake components must also be checked.
3   Once you are into the red zone, adjusting the brakes will cease to be effective.
4   Replace the shoes then re-adjust the indicator on the serrations to read at the extreme left of the usable range plate in rear brake shoe replacements and to the extreme right of the plate in the case of front brake lining replacements.

14 Disc brakes - general

The section on disc brakes applies only to S2 and S3 models from 1973 onwards. The discs and pads which carry out the braking operation in this system are open to direct contact with the air stream when the machine is in motion. This gives first class dissipation of the heat generated in braking and minimises the danger of brake fade due to overheating common to drum brakes.

15 Master cylinder - examination and renovation

1   The master cylinder and master cylinder body are secured by the master cylinder mounting to the right hand side of the handlebars. It is unlikely to give trouble unless the machine has been unused for a long period or a considerable mileage has been recorded. Indications of wear are brakes not holding, poor pressure, leakage of fluid and a gradual fall in the level of the fluid in the cylinder. Keep an eye on the level line in the cylinder.
2   Keep the relief port hole in the bottom of the reservoir clear or brake drag may result.
3   Repairs to the master cylinder require dismantling and reassembly with a high degree of skill and cleaner than normal conditions. It is recommended that you obtain a replacement unit of the correct type in lieu of repair.
4   To remove the cylinder drain out the fluid as detailed in Section 22 (Fluid changing) of this Chapter. Detach the plastic tube.
5   Remove the hose, the handlebar lever pivot bolt and the lever itself.
6   It is now possible to remove the piston assembly together with the rubber seals.
7   Keep a careful note of the seal arrangement as they must be replaced in the same order to avoid brake failure.
8   Examine the components for wear or scratching and the seals for edging failure or rubber deterioration. Renew if evident.

12.3 Length of travel is controlled by the adjusting nut at the end of the brake operating lever

9   Soak new seals in clean fluid for at least 15 minutes prior to refit.
10 Apply brake fluid to the inner wall of the cylinder.
11 Ensure the primary cup and check valve are not installed backwards or not turned sideways after insertion.
12 Fit a new retaining ring. You may need a special tool to fit it into the cylinder wall.
13 The same tool can be used for fitting the boot and boot stop.
14 Refit the master cylinder in reverse order to that of the dismantling. Reconnect the handlebar lever, hose etc.
15 Refill the system with new fluid as specified in the fluid section of this Chapter and bleed as described in Section 21 of this Chapter.
16 Ensure the brake is working correctly before giving a road test. Use the brake gently for a day or so to restore pressure and align the pads correctly.
17 The front brake adjusts itself automatically during use and if any further adjustment is necessary it is usually due to worn parts or faults.
18 If the front brake lever vibrates, adjust the lever leaving a certain amount of play for a good braking movement.
19 The lever is adjusted by loosening the locknut, turning the adjusting bolt a part of a turn to give the lever not more than 3/16 inch (5 mm) play. Tighten the locknut.
20 Kawasaki tools are recommended for use when removing the seals, boot and boot stop, although the experienced owner will probably already possess the equivalents. These are:
a) Seal hook — Kawasaki part no 56019-111
b) Seal installing tool — Kawasaki part no 56019-109
c) Boot and boot stop installation tool — Kawasaki part no 56019-110.

16 Front wheel with disc brake - removal and dismantling

1   Disconnect the speedometer cable.
2   Remove the spindle clamp bolts.
3   Lift and support the machine so that the wheel can be lifted out.
4   Take great care when dismantling the front hub to take a firm hold of the speedometer gearbox (not the spindle) and unscrew the spindle.
5   Remove the wheel cap and collar.
6   Tap evenly around the inner race of the right bearing from the left side of the wheel and knock it out.
7   Remove the distance collar and then the oil seal. The oil seal must be replaced.
8   Remove the retaining ring and now from the right side of the wheel, tap evenly around the inner race of the left bearing and knock it out.

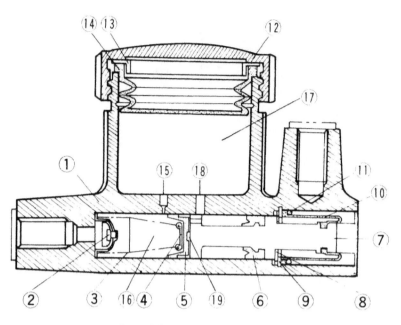

Fig. 5.3. Disc brake master cylinder

1	Master cylinder body	6	Secondary cup	11	Stopper, dust seal	16	Pressure chamber
2	Check valve	7	Piston	12	Cap	17	Reservoir
3	Spring	8	Stopper, piston	13	Plate	18	Supply port
4	Spring seat	9	Circlip	14	Cap seal	19	Non-return valve
5	Primary cup	10	Dust seal	15	Relief port		

17 Removing the caliper unit

1   Remove the front wheel as described in the previous Section.
2   Remove the brake pads. They will slip out of position without need to separate the caliper unit.
3   Unscrew the brake pipe nipple and disconnect the pipe.
4   Use a rubber bleeder valve cap over the end of the pipe to save loss of fluid.
5   Loosen the socket head shafts through the calipers. These screws are normally very tight.
6   Remove the two mounting bolts and remove the caliper assembly from the wheel.
7   Carefully remove the socket head screws and remove the right hand caliper and its associated friction pad.
8   Remove the caliper holder from the shafts and take out the other friction pad.
9   Do not damage the O rings or the boots.
10 Remove the piston dust seal and gently remove the piston from the left hand caliper.
11 Examine the rubber hose in the brake line connections. If cracks or swellings have appeared then replace the hose length.
12 Examine the plated steel tube to the caliper nipple. Damage is often found in the form of rust.
13 Too much emphasis cannot be made on the examination of the fittings on the steel tube. The fit should be perfect and tightly screwed into place.
14 If the caliper unit is separated, the shafts must be renewed.

18 Overhaul and replacement of caliper unit and friction pads

1   The friction pads should be examined for wear. The correct type carries a red line to indicate wear limit. Replace both pads as a set.
2   Grease or oil on the pads can be washed off with petrol but if any doubt of their condition, renew both as a pair.

3   Clean the recesses into which the pads fit and the exposed end of the pistons which actuate the left hand pad. Use only a small soft brush. Smear the piston face and the brake pad recesses with hydraulic fluid to act as a lubricant.
4   The oil seal around the piston should be examined for wear or damage. Replace if in any doubt.
5   Examine the cylinder and piston for rust or damage and replace if necessary.
6   Replace the oil seal every other time the pads are changed but change it if there is a large difference in wear between left and right friction pads.
7   Caliper standards are as follows:

Cylinder inside diameter 1.5031 - 1.5039 inch
Replace at 1.5045 inch (38.25 mm)

Piston outside diameter 1.5006 - 1.5019 inch
Replace at 1.5002 inch (38.105 mm)

8   Replace the caliper and friction pads in the order or removal.
9   It is now necessary to bleed the brake as described in Section 21 of this Chapter.

19 Removal and overhaul of disc unit

1   The disc should not be omitted when servicing the brake system.
2   Oil on the disc can be the cause of poor braking. It can be cleaned off with petrol.
3   The disc on a machine can warp and cause the brake pads to drag and wear themselves and the disc prematurely. This is known as dragging and will cause overheating and poor braking.
4   Discs can be measured for wear preferably with a thickness

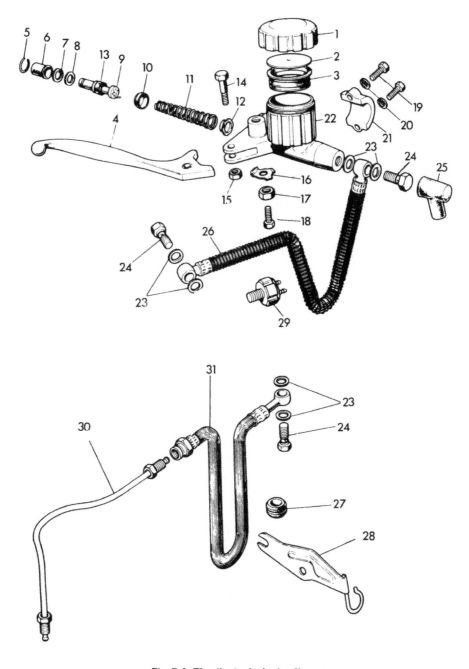

Fig. 5.4. The disc brake hydraulic system

1	Cap	9	Piston assembly	17	Nut	25	Dust cover
2	Plate	10	Primary cup	18	Bolt	26	Hose
3	Cap seal	11	Spring assembly	19	Bolt	27	Grommet
4	Brake lever	12	Check valve assembly	20	Washer	28	Bracket
5	Dust seal stopper	13	Secondary cup	21	Master cylinder mounting	29	Pressure switch
6	Dust seal	14	Bolt	22	Master cylinder body	30	Brake pipe
7	Circlip	15	Nut	23	Washer	31	Hose
8	Piston stopper	16	Lock washer	24	Banjo bolt		

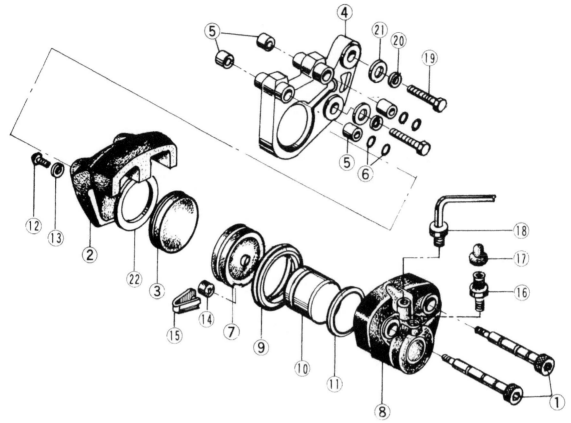

Fig. 5.5. Front disc brake and caliper

1	Allen-head shaft	7	Pad A	13	Lock washer	19	Mounting bolt
2	Caliper B	8	Caliper A	14	Bushing	20	Lock washer
3	Pad B	9	Piston dust seal	15	Stopper	21	Washer
4	Caliper holder	10	Piston	16	Bleeder valve	22	Ring
5	Boots	11	Seal	17	Bleeder valve cap		
6	'O' ring	12	Screw	18	Nipple		

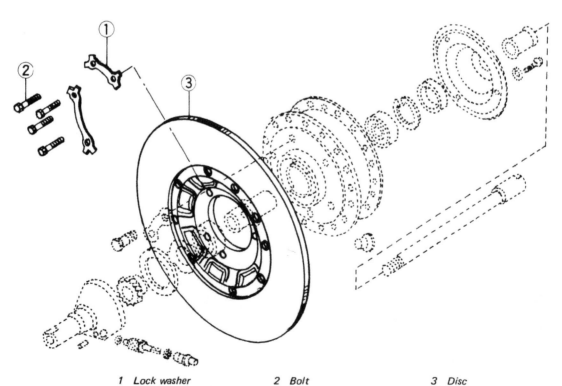

1   Lock washer   2   Bolt   3   Disc

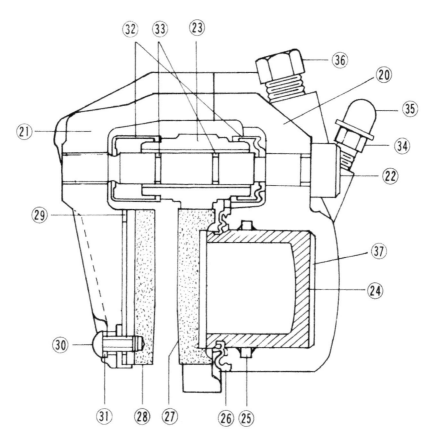

Fig. 5.6. Disc brake caliper unit

20 Caliper A	25 Oil seal	30 Screw	35 Bleeder valve cap
21 Caliper B	26 Dust seal	31 Lock washer	36 Nipple
22 Allen-head shaft	27 Pad A	32 Dust seal	37 Cylinder
23 Caliper holder	28 Pad B	33 'O' ring	
24 Piston	29 Ring	34 Bleeder valve	

gauge. The standard thickness should be 0.276 inch (7 mm). If it is worn down to below 0.217 inch (5.5 mm) at any point it should be replaced.

5 The run-out on the disc should be under 0.004 inch (0.1 mm). Replace if it exceeds 0.012 inch (0.3 mm).

6 To remove the disc the wheel should first be removed from the machine.

7 Straighten up the mounting bolt lock washers and remove the mounting bolts, then lift off the disc.

8 When refitting the disc bolts tighten them to a torque setting of 140 - 190 inch pounds (1.6 - 2.2 kg m).

20 Front wheel with disc brake - assembly

1 Assembly of the front hub of a wheel fitted with a disc brake is the reverse of the dismantling procedure. Points to watch are as follows:

2 Renew the oil seal. You may need a special tool for replacing it, and the bearings.

3 Hold the gearbox stationary when the spindle is screwed in, not the spindle.

4 When replacing the spindle clamps first tighten the front spindle clamp bolt and then the rear bolt each side. There should be a gap at the rear only, after tightening.

5 Tighten to 13 - 14.5 ft lbs (1.8 - 2.0 kg m) torque setting.

6 When refitting the speedometer cable turn the wheel so that the speedometer drive shaft tongue will mate with the cable end groove.

21 Bleeding the front disc brake

1 If air enters the brake lines, brake lever movement will be lost in compressing the air. This in turn loses braking power and causes what is commonly known as 'spongy' brakes.

2 First check that the fluid in the reservoir is at least up to the indicated level line. During the exercise replenish as necessary. If the reservoir empties during the operation commence again from the beginning with a full reservoir.

3 With the reservoir cap removed, gently pump the brake lever until no air bubbles can be seen rising through the fluid from the holes at the bottom of the reservoir.

4 Replace the reservoir cap.

5 Connect a clear plastic pipe to the caliper bleed valve. The other end of the pipe should run into a container so that the tube end is immersed in brake fluid. Open the bleeder valve on the caliper unit by turning it about a half turn. The end of the plastic pipe must, at all times, be submerged completely in clean brake fluid within the container.

6 Gently pump the brake lever until it feels hard and with the lever still gripped tightly with the left hand quickly close the

bleeder valve.

7 Release the lever.

8 Repeat the operation until no more air can be seen escaping along the plastic pipe. Remove the plastic pipe and refit the rubber cap to the bleeder valve.

9 Top up the reservoir to the indicated line.

10 If a double disc brake is fitted a similar exercise must be carried out on the other side of the wheel.

11 Always use clean brake fluid and never the fluid expelled into the container unless it has stood for at least twenty four hours so that all the air bubbles have dispersed.

22 Disc brakes - fluid changing and precautions

1 All motor vehicle manufacturers emphasise the precautions to be taken when refilling or renewing the brake fluid. These are so numerous that they have been formed into the following list. With any new machine it is essential to follow the manufacturer's recommendations in your handbook and use the manufacturer's recommended fluid. These are listed in the Specifications Section at the beginning of this Chapter.

2 Do not re-use the old brake fluid from the machine after draining the system.

3 If the fluid has been left in a can with the cap off or has been in use for topping up for a long time, replace it with a new tin and dispose of the old one.

4 Do not mix different types of fluid. This can cause ineffective braking and deterioration of the brake rubber seals.

5 Keep to the same brake fluid when topping up. (It may be difficult to ascertain what is already in the system.)

6 The correct fluid should be in a can marked DOT3. A typical can used in the UK has a plastic filler spout and claims to exceed SAE J1703b and the US Federal Motor Vehicle Safety Standard 116 DOT 3. It is suitable for systems requiring a fluid boiling point of 550°F. It holds 8.8 fluid ounces or 0.25 litres. It is possible that European countries will operate their own equivalent specification.

7 Do not leave the reservoir cap off as moisture can be absorbed especially in the rain.

8 Disc brake components can be cleaned with disc brake fluid which is the most available fluid. If fluids such as ethyl alcohol are used for cleaning components they should not be in contact with the fluid for longer than 30 seconds.

9 Protect disc brake parts from petrol or oil.

10 If you accidentally loosen the bleed valve or any of the fittings at any time, rebleed the system.

11 Application of the brakes causes heat to be generated by friction between the disc and the pads. Some of the heat is dissipated in the brake fluid and could cause its temperature to rise to 300°F (150°C). Unless the correct brake fluid is used and the fluid changed once a year or every 6000 miles (whichever is earlier) the brake fluid becomes contaminated and could boil at a lower temperature and cause a vapour lock in the line.

12 To gain access to the master cylinder first attach a bleed tube to the caliper bleed valve. Bleed into an old container or jar. Open the bleed valve and pump the brake lever until no fluid remains.

13 Close the bleeder valve and fill with the new fluid recommended by the manufacturer. Keep the bleed tube on the valve, immersed in brake fluid.

14 Re-open the bleeder valve, squeeze the brake lever and close the bleeder valve still squeezing the lever. Quickly release the lever.

15 Repeat the movement until fluid starts squirting from the bleeder hose.

16 Check that the fluid in the reservoir is kept up to the level line. It is necessary to refill if much fluid is ejected into the container.

23 Rear wheel sprocket - removal, examination and replacement

1 The rear wheel can be removed from the machine by following

the instructions given in Section 8 of this Chapter.

2 If it is found necessary to remove the rear sprocket and coupling:

3 Remove the left hand chain adjuster.

4 Remove the wheel bearing and oil seal.

5 Remove the tab washers, the four retaining nuts, the sprocket itself and then the spindle sleeve.

6 Examine the sprocket teeth for wear or damage. If damage is discovered it is expedient to change front and rear sprocket and the chain to avoid further damage to the set.

7 There is no advantage in varying the size of either sprocket. The sizes selected have been chosen by the manufacturer as the result of exhaustive tests, to give optimum performance with the existing engine characteristics.

8 If sprockets are changed note that the S3 model (400 cc) rear wheel sprocket has 41 teeth whilst that of the S2 model (350 cc) has 43 teeth. Do not confuse the sizes.

9 Always use new tab washers when fitting sprocket retaining nuts.

10 If the sprocket is bent, rapid chain wear will occur and chain adjustment be made difficult.

11 A simple test with the rear sprocket removed is to lay it on a flat surface and check the gap between the sprocket and the surface at various points. If it is greater than 0.02 inch at any point it is expedient to replace the sprocket.

12 To check for sprocket wear take accurate measurements from the base of a suspected tooth across to the base of the directly opposite tooth. Sprockets should only have to be changed for wear after a fair mileage. If the sprocket is wearing rapidly, check the chain for wear.

13 Rear sprocket dimensions:

Machine Model	Teeth	Diameter at base of teeth	
		Normal	Replace at
S1	48	9.16 in (232.6 mm)	9.07 in (230.5 mm)
S2	43	8.16 in (207.3 mm)	8.09 in (205.5 mm)
S3	41	7.76 in (197.2 mm)	7.70 in (195.5 mm)

14 Reassembly of the rear sprocket and coupling to the rear wheel is the reverse procedure to dismantling.

24 Chain - examination, lubrication and replacement

1 The chain is fully exposed apart from the protection of a chainguard along the upper run and if not properly maintained will have a short life. Together with the front and rear sprockets it is a means of providing secondary reduction. A worn chain will cause rapid wear of the sprockets and they too would need replacing.

2 The chain is removed or replaced by means of uncoupling the master link. Slacken off the chain adjusters for the operation.

3 Always ensure that the link is refitted in the correct direction. The closed end of the clip should face in the direction of rotation of the chain, otherwise the clip could become disengaged.

4 A new chain will stretch by a relatively large amount at first and this must be checked more frequently for the correct tension than a chain that has been well used.

5 Chain wear occurs between the pin and bush and also the bush and roller due to chain movement and tension, and the chain will stretch. The constant friction between chain roller and sprocket can also cause slackness.

6 To check the chain for wear, first clean off the old oil and grease and lay the chain lengthwise in a straight line before compressing it endwise. Fix one end and pull the chain out in the opposite direction to see the extent of play. If greater than ¼ in per foot replace the chain. An indication of chain wear is the amount a chain can be bent sideways.

7 The chain can be cleaned in petrol and lubricated with SAE 30 oil but it is preferable to use Linklyfe or Chainguard for better lubrication. Remove the chain and immerse it in the molten lubricant recommended, such as Chainguard, after it has been cleaned in a paraffin bath. These lubricants achieve better

23.3. Remove the chain adjuster

23.4 Remove the ball bearing and oil seal

23.5 And then the spindle sleeve

24.3 Always ensure the link is fitted in the right direction

24.10 Set to the same mark on the swinging arm fork

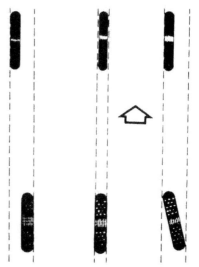

Fig. 5.7. Checking wheel alignment

penetration of the chain links.

8   If the chain has become beyond adjustment it is possible
your dealer may use a chain rivet extractor or chain breaker to
shorten it by one link. The normal number of links should be
S1 - 102, S2 - 98, S3 - 104. This chain shortening however is not
advised unless carried out by your main dealer.

9   Chain play can be checked more accurately with the chain
adjusted as far as possible on the machine with the machine on
level and firm ground. Check chain play at the centre of the
chain. It should be greater than 3/8 inch (10 mm) and must be
replaced if greater than 1½ inches (40 mm). The recommended
amount of play is 3/4 inch (20 mm).

10 To keep the wheels and chain aligned it is essential to have
both chain adjusters set to the same mark on the swinging arm
fork. If in doubt, check by placing a plank along both wheels, as
shown in the accompanying diagram.

11 To tighten the chain screw in the chain adjuster bolts.

12 Loosen the chain by screwing out the bolts and pushing the
rear wheel forward.

25 Tyres - removal and replacement

1   Tyre changing will be required either due to wear or a puncture. To repair a puncture follow the instructions of the manufacturers of the repair outfit. It is however good practice to renew
an inner tube when it becomes punctured.

2   Remove the wheel as described in this Chapter; see Section 2
for the front wheel, or 8 for the rear wheel.

3   Deflate the tyre by removing the Schrader valve and push both
the beads of the tyre away from the wheel rim and into the well.
Remove the locking ring of the inner tube valve and push the
valve right into the tyre.

4   Insert a tyre lever close to the valve and lever the edge of the
tyre over the outside of the wheel rim. Very little force should
be necessary; if resistance is encountered it is probably due to
the fact that the tyre beads have not entered the well of the wheel
rim all the way round the tyre.

5   Once the tyre has been edged over the wheel rim, it is easy to
work around the wheel rim so that the tyre is completely free on
one side. At this stage, the inner tube can be removed.

6   Working from the other side of the wheel, ease the other edge
of the tyre over the outside of the wheel rim that is furthest away.
Continue to work around the rim until the tyre is free completely
from the rim.

7   If a puncture has necessitated the removal of the tyre, re-
inflate the inner tube and immerse it in a bowl of water to trace
the source of the leak. Mark its position and deflate the tube. Dry
the tube and clean the area around the puncture with a petrol-
soaked rag. When the surface has dried, apply the rubber solution
and allow this to dry before removing the backing from the patch
and applying the patch to the surface.

8   It is best to use a patch of the self-vulcanising type, which will
form a very permanent repair. Note that it may be necessary to
remove a protective covering from the top surface of the patch,
after it has sealed in position. Inner tubes made from synthetic
rubber may require a special type of patch and adhesive, if a satisfactory bond is to be achieved.

9   Before replacing the tyre, check the inside to make sure the
agent that caused the puncture is not trapped. Check also the
outside of the tyre, particularly the tread area, to make sure
nothing is trapped that may cause a further puncture.

10 If the inner tube has been patched on a number of past
occasions, or if there is a tear or large hole, it is preferable to discard it and fit a replacement. Sudden deflation may cause an
accident, particularly if it occurs with the front wheel.

11 To replace the tyre, inflate the inner tube sufficiently for it to
assume a circular shape but only just. Then push it onto the tyre
so that it is enclosed completely. Lay the tyre on the wheel at an
angle and insert the valve through the rim tape and the hole in the
wheel rim. Attach the locking cap on the first few threads, sufficier
to hold the valve captive in its correct location.

12 Starting at the point furthest from the valve, push the tyre

Fig. 5.8a Tyre removal

A   Deflate inner tube and insert lever in close proximity to tyre valve
B   Use two levers to work bead over the edge of the rim
C   When first bead is clear, remove tyre as shown

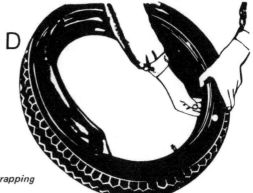

Fig. 5.8b. Tyre fitting

D *Inflate inner tube and insert in tyre*
E *Lay tyre on rim and feed valve through hole in rim*
F *Work first bead over rim, using lever in final section*
G *Use similar technique for second bead. Finish at tyre valve position*
H *Push valve and tube up into tyre when fitting final section, to avoid trapping*

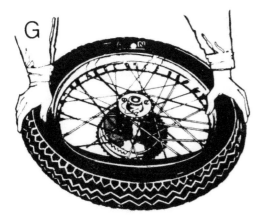

bead over the edge of the wheel rim until it is located in the central well. Continue to work around the tyre in this fashion until the whole of one side of the tyre is on the rim. It may be necessary to use a tyre lever during the final stages.

13 Make sure there is no pull on the tyre valve and again commencing with the area furthest from the valve, ease the other bead of the tyre over the edge of the rim. Finish with the area close to the valve, pushing the valve up into the tyre until the locking cap touches the rim. This will ensure the inner tube is not trapped when the last section of the bead is edged over the rim with a tyre lever.

14 Check that the inner tube is not trapped at any point. Re-inflate the inner tube, and check that the tyre is seating correctly around the wheel rim. There should be a thin rib moulded around the wall of the tyre on both sides, which should be equidistant from the wheel rim at all points. If the tyre is unevenly located on the rim, try bouncing the wheel when the tyre is at the recommended pressure. It is probable that one of the beads has not pulled clear of the centre well.

15 Always run the tyres at the recommended pressures and never under or over-inflate. The correct pressures for solo use are shown in the Specifications Section of this Chapter. If a pillion passenger is carried, increase the rear tyre pressure as indicated.

16 Tyre replacement is aided by dusting the side walls, particularly in the vicinity of the beads, with a liberal coating of French chalk. Washing-up liquid can also be used to good effect, but this has the disadvantage of causing the inner surfaces of the wheel rim to rust.

17 Never replace the inner tube and tyre without the rim tape in position. If this precaution is overlooked there is good chance of the ends of the spoke nipples chafing the inner tube and causing a crop of punctures.

18 Never fit a tyre that has a damaged tread or side walls. Apart from the legal aspects, there is a very great risk of a blow-out, which can have serious consequences on any two-wheel vehicle.

19 Tyre valves rarely give trouble, but it is always advisable to check whether the valve itself is leaking before removing the tyre. Do not forget to fit the dust cap, which forms an effective second seal.

20 Examine the tyres at regular intervals for stones, side wall damage or bulges. Remove the stones and re-use but side wall damage or bulges will mean a replacement tyre.

26 Fault diagnosis - wheels, brakes and tyres

Symptom	Cause	Remedy
Handlebars oscillate at low speeds	Buckle or flat in wheel rim, most probably front wheel	Check rim alignment by spinning wheel. Correct by retensioning spokes or having wheel rebuilt on new rim.
	Tyre not straight on rim	Check tyre alignment.
Machine lacks power and accelerates poorly	Brakes binding	Warm brake drums provide best evidence. Re-adjust brakes.
Brakes grab when applied gently	Ends of brake shoes not chamfered	Chamfer with file.
	Elliptical brake drum	Lightly skim in lathe (specialist attention needed).
Brake pull-off sluggish	Brake cam binding in housing	Free and grease.
	Weak brake shoe springs	Replace, if brake springs not displaced.
Harsh transmission	Worn or badly adjusted chains	Adjust or replace as necessary.
	Hooked or badly worn sprockets	Replace as a pair, together with chain.
Steering pulls to one side	Rear wheel misalignment	Adjust rear wheel alignment.

Chapter 6 Electrical system

Contents

Specifications

Battery
Type	Lead acid
Make	Furukawa Batt. Co 12N5 or Yuasa 12N113B
Voltage	12 volts
Capacity (ten hour rate)	5.4 ampere hours/5.5 ah
Earth	Negative

Generator (AC)
Type	Kokusan AR2101
Output	12 volt min 16 volt max 5.3 amperes @ 1500 rpm

Main fuse ... ... — 20 amperes (spare carried)

Bulbs
Main headlamp	12v 35/35W (S1B, S3 European) 35/25W other models
Pilot lamp	12v 4W (S1, S2, S1B, S3 European) USA not fitted
Tail/stop lamp	S1B, S3 (European) 12v 5/21W; S1, S2, S1B, S3 12v 8/23W
Speedometer lamp	12v 3W S1, S2, 2 fitted; others 1 only
Tachometer lamp	12v 3W S1, S2, 2 fitted; others 1 only
Neutral indicator lamp (1)	12v 3W
High beam indicator lamp (1)	12v 1.5W
Flashing indicator lamps (4)	12v 21W (S1B - S3 European) 12v 25W (others)

Regulator ... ... — Kokusan RS 2114

Flasher unit ... ... — 12v (23 x 2 + 3.4W)

Ignition coils ... ... — Diamond TU-29M-14 (3)

Rectifier ... ... — Full wave - silicon diodes

---

1 General description

The Kawasaki triple cylinder machines covered by this manual are fitted with a crankshaft driven alternator and a 12 volt lead acid accumulator. The DC system uses a negative earth return.

The output of the alternator is three phased and it is necessary to rectify this voltage before charging the battery.

The merit of the AC generator over the DC type fitted to motor cars is its compactness, light weight and most of all lack of parts such as brushes liable to wear and failure. The 'S' type machines normally have a permanent magnet type field rotor for the alternator.

As repeated several times in this manual, the use of a mini-

multimeter is recommended if tackling electrical problems. The
average motorcyclist expects to purchase from time to time
several extremely useful tools such as impact screwdrivers and
torque wrenches. A small multimeter is often cheaper to purchase
than one of these.

Lighting systems vary slightly between models, and countries
sold to, due to legislation in these countries. As an example most
'S' series models have a front brake stop lamp switch which is not
fitted to models supplied to the UK. Where possible, all alternatives are given.

2 Battery - in situ maintenance

1   The battery is a 12v Furukawa Battery Co type number 12N5
and has a 5.4 ampere hour capacity. The vent tube hangs down
through the frame.

2.1 The battery is a 12V Furukawa

2   The transparent case of the battery allows the upper and
lower levels of the electrolyte to be observed when the dualseat
is lifted. Maintenance is normally limited to keeping the electrolyte level between the upper and lower limits and making sure the
vent tube is not blocked. The lead plates and their separators can
be seen through the transparent case. A further guide to the condition of the battery, ie a charged battery has plates of a muddy
brown colour. Grey plates indicate a high level of sulphation and
the battery should be renewed. If the battery case contains sediment at the bottom, the plates are disintegrating and the battery
will soon need replacing.
3   Unless acid is spilled (as may occur if the machine falls over),
the electrolyte should always be topped up with distilled water,
to restore the correct level. If acid is spilled on any part of the
machine, it should be neutralised with an alkali such as washing
soda and washed away with plenty of water, otherwise serious
corrosion will occur and weakening of the affected structure.
Keep acid off the hands and clothing and wash quickly if an
accident occurs with the battery to cause leakage. An acid splash
in the eyes calls for immediate first aid treatment.
4   Top up with sulphuric acid (from a garage) of the correct
specific gravity (1.260 - 1.280) only when spillage occurs. Most
good class filling stations have distilled water available on the
forecourt and often a hydrometer for checking SG. Always check
that the battery vent pipe is well clear of the frame tubes or any
of the other machine parts.
5   Do not try to repair a battery, not even a specialist would
try! Renew it as soon as possible. Normal battery life is two to
three years.
6   If your machine has been rested for a period of some weeks
give it a refresher charge from a battery charger. If the battery is
permitted to discharge completely, the plates will sulphate and
render the battery useless. Do not ever allow the electrolyte to
fall below the top of the plates.

3 Battery charging procedure

1   The normal charging rate for the 5.4 ampere hour battery is
0.54 amperes for ten hours. This is for a 100 per cent efficient
battery. In an emergency with a very flat battery this charge can
be increased to 0.8 and 1.2 amps respectively but this is not
recommended as it can cause shedding of the active material.
Your battery charger will need to have a variable resistor to
enable you to adjust the charge to 0.55 amps. Remove the vent
plugs when charging and ensure the charger red lead goes to
positive and the black lead to negative.
2   When refitting the battery into its holder examine the holders
for any signs of spilled acid if any are seen take action as recommended in Section 3.3. Make sure you reconnect your leads
correctly and with everything switched off. Always apply a thin
coating of Vaseline to the terminals before reconnecting.
3   Do not smoke or cause an open flame (such as using a blow
lamp for welding) near a charging battery. Highly explosive
hydrogen gas is given off during the charging chemical process.
You must take a similar precaution with petrol vapour.

4 Crankshaft driven alternator - checking the output

1   Step by step operations for removing the alternator have been
given in Chapter 1 of this manual.
2   The crankshaft of the machine drives a magnetised rotor within a set of stationary coils called a stator and by magnetic induction causes a current to occur in the stator. By increasing the
number of coils or windings on the coils this current can become
comparatively large and if rectified can be used for charging the
battery and operating the electrical accessories.
3   The rotor on the machine illustrated is a permanent magnet
type and its output is controlled by a full wave selenium rectifier.
4   If you are having electrical troubles such as dull lights and a
battery showing signs of not being charged as indicated by the
colour of the plates do not immediately suspect your alternator.
It should last the life of the machine. Measure the volts across the
12 volt battery with the machine running at a normal cruising rate.
It should show up to 15 volts DC. If it does not then thoroughly
check the connections and suspect the alternator, regulator or
rectifier. At night there should be a significant improvement in
lighting indication (if your battery has deteriorated or flattened)
when the engine is running at a normal cruising speed to indicate
that the alternator is serviceable.
5   Failure to read 14.5 - 15 volts across the battery with the
engine running normally indicates electrical trouble. Suspect
faulty connections or damaged cables. If you have a multimeter
switch to AC volts and measure the AC input to the rectifier. It
should be about 15.5 volts AC. If this AC voltage is present then
switch back to DC volts and measure the output from the rectifier.
You will have to disconnect cables to carry out these checks so
refer to the circuit diagram for your machine.

5 Rectifier location

1   The function of the full wave selenium rectifier is to convert
the AC current produced by the alternator to DC so that it can
be used to charge the battery. The rectifier is the full wave type
which prevents discharge of the battery through the alternator
with a stopped engine or if the output voltage of the alternator
falls below that of the battery.
2   The rectifier with other electrical components is beneath the
seat and protected by a removable cover where it is not directly
exposed to water or battery acid which may damage it.
3   The rectifier is unlikely to cause trouble if normally operated
and cannot be repaired - only replaced.
4   The rectifier can be damaged if you run your machine without
a battery for any length of time. A high voltage will develop in
the absence of any load on the electrical coils, which will cause
a reverse flow of current and consequent damage to the rectifier
cells. A reverse connection of the battery will similarly cause

damage. If you haven't the test equipment necessary do not tamper with the rectifier, it may not be faulty and you could damage it. It should be checked by a Kawasaki agent or an auto-electrical expert, who has the necessary test equipment.

fraying - you may even smell burnt insulation. It might be cheaper to wheel the machine home or contact the AA or nearest garage.

6 Fuse - location and replacement

1   A fuse is incorporated in the electrical system to give pro-tection from a sudden overload as may occur during a short circuit. It is found within a fuse holder which forms part of the wiring snap connections and is clipped to the battery holder. A transparent plastic bag attached to the wiring carries a spare fuse for use in an emergency. The fuse is rated at 20 amps.
2   If a fuse blows, it should be replaced, after checking, to ensure that no obvious short circuit has occurred. If the newly fitted fuse of the correct rating then blows the circuit will need a thorough examination. Do not panic when the first fuse blows as fuses like humans become tired with age and exertion and fail.
3   When a fuse blows and the machine is running and you have no spare available a get-you-home-once only routine is to wrap the spent fuse in silver paper and replace in the holder. The silver paper will act as a conductor of electricity. Replace it without fail at the first opportunity. This expedient is not to be carried out if there is evidence of a malfunction. Look for burning or

7 Regulator unit location

1   The regulator unit associated with the rectifier is located adjacent to the flasher relay unit and the rectifier and all are close to the air filter unit.
2   A regulator is needed to control the output from the crank-shaft driven alternator. This is due to the fact that voltage is derived from the permanent magnet rotor field cutting and inducing a current in the stator. The faster it rotates the more it cuts and the higher the voltage produced. If this high voltage is not controlled or regulated it could cause electrical burn-outs.
3   The devices used to regulate are also semi-conductors called thyristors which can be switched on instantly if the voltage exceeds 16 volts and bring in a resistive load to reduce the volt-age. They switch off this load the instant the voltage drops to 15.5 volts.
4   Unless you are an electrical expert do not meddle with this section of your motorcycle. It is in a sealed container and must be replaced if faulty.

5.2a The rectifier is located in a position not directly exposed to water etc.

5.2b The rectifier is unlikely to cause trouble

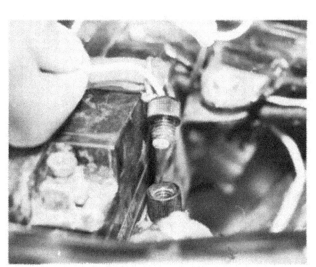

6.1 A fuse is incorporated in the electrical system

7.1 The regulator unit is located adjacent to the flasher relay unit

8 Headlamp - examination and bulb replacement

1   Always keep the headlamp in a clean condition. Dust and mud at night can cause loss of light intensity if covering the lens. The unit is mounted by two bolts which can be loosened to permit correct level adjustment.

2   To replace the dual filament bulb in the event of failure of the head or high beam light first remove the two screws at the back of the lamp with one hand whilst holding the rim unit with the other. The lamp and holder are push-in and can be lifted from the reflector. The lamp is of the bayonet fitting type but will only fit in one way due to the staggering of the side contacts, ie offset.

3   The pilot lamp bulb is also of the bayonet fitting type and fits within a bulb holder which has the same form of connection to the headlamp connector. This lamp has a 3w rating.

4   Check the connections within the headlamp are correctly made before refitting the lamp assembly especially if the handlebars have been changed.

5   The high beam indicator lamp fits into the headlamp shell. To remove it, slide back the rubber sleeve and pull out the holder and the lamp bulb.

6   On some 'S' series models the high beam indicator is located between the speedometer and tachometer between two other indicator lamps the top of which is the flasher indicator. It is marked BE and shows a blue light.

7   The beam height of the headlamp is adjusted by slackening the mounting bolts and tilting either way.

8   The machine should be placed on level ground about twenty five feet away from a wall in its normal position - not on its stand and with rider and pillion passenger mounted. On main beam the height of the centre of the reflected light from the ground should be the same as the bulb from the ground. This will ensure that when the dipswitch is operated the lamp will not dazzle the oncoming motorist or cyclist.

8.2a Remove the two screws at the back of the lamp to free the reflector unit

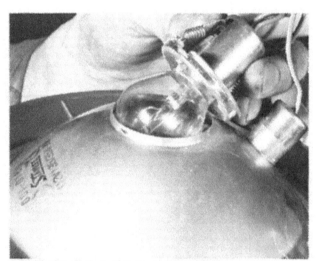

8.2b The headlamp bulb is bayonet type fitting

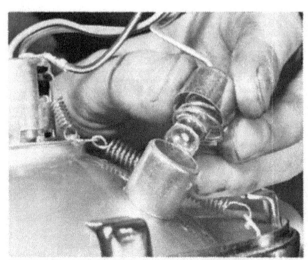

8.3 The pilot lamp is within the headlamp assembly

8.5 The high beam indicator lamp fits into the headlamp shell

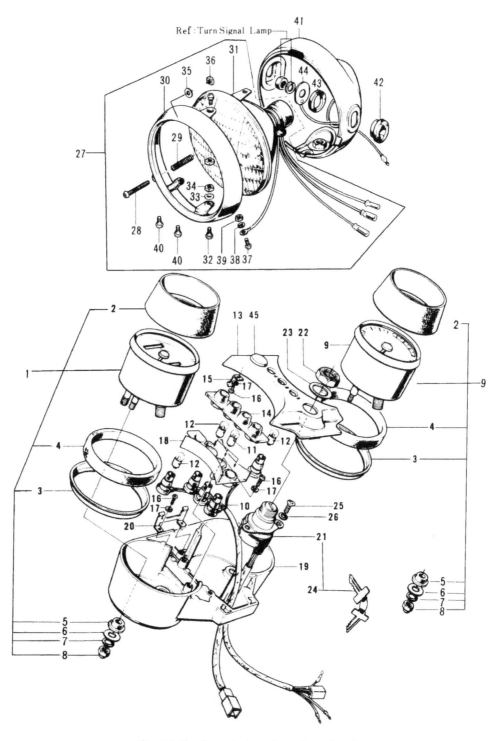

Fig. 6.1. Headlamp, tachometer and speedometer

1	Speedometer head	13	Indicator bulb cover
2	Cover - 2 off	14	Anti-vibration mounting
3	Meter ring - 2 off	15	Cover holder - 2 off
4	Meter damper rubber - 2 off	16	Screw - 6 off
5	Anti-vibration rubber - 4 off	17	Spring washer - 6 off
6	Plain washer - 4 off	18	Bulbholder cover
7	Spring washer - 4 off	19	Meter case
8	Nut - 4 off	20	Meter case retainer
9	Tachometer head	21	Ignition switch
10	Bulbholder socket - 5 off	22	Ignition switch nut
11	Bulb, 12V, 1.5W	23	Ignition switch washer
12	Bulb, 12V, 3W - 5 off	24	Ignition key set

25	Screw - 2 off	37	Screw
26	Spring washer - 2 off	38	Spring washer
27	Headlamp unit	39	Plain washer
28	Focus adjusting screw	40	Screw - 2 off
29	Focus adjusting spring	41	Headlamp shell
30	Headlamp rim	42	Anti-vibration mounting - 2 off
31	Sealed beam unit	43	Anti-vibration mounting - 2 off
32	Screw - 2 off	44	Collar set - 2 off
33	Plain washer - 2 off	45	Rubber plug
34	Nut - 2 off		
35	Washer - 2 off		
36	Nut		

9 Stop and tail lamp - bulb replacement

1   The stop and tail lamp is replaced by removing the four cross-head screws securing the lens to the assembly. Hold the lens securely in one hand as it is plastic and expensive to replace and remove the bulb which is offset bayonet fitting. The bulb is 12v 23/3 watts. One 23 watt filament is for the stop warning lamp and the 3 watt one is for the normal tail light.

2   The stop warning lamp is operated by a switch attached to the rear brake pedal mechanism. One end of the switch is connected to a cable connected to the rear brake light and the other end is moveable and operated by a spring attached to the brake pedal lever. Ensure the spring is in place if refitting.

10 Flashing indicators - bulbs and flasher relay location and replacement

1   The four flashing indicator lamps are each fitted with 12 volt 21 watt lamps and are mounted on stalks through which the cables are fed.

2   To replace a lamp remove the plastic lens cover by removing the two retaining crosshead screws. Push the bulb in, turn to the left and withdraw.

3   The flasher relay unit is located under the dualseat close to the battery assembly. It is rubber mounted to protect it from vibration. This unit is fairly easily replaced but suspect first the four bulbs affected (by swapping around), the cable connections and most of all the operating switch on the handlebars.

4   Listen for audible clicks to ensure the flasher unit is operating.

5   Handle the unit with care and do not drop it as it is very easily damaged.

6   Check the rating of the bulb if you need to replace one and be sure to fit the correct replacement. The legal requirement in the UK is 60 - 120 flashes per minute and this can be varied if you fit a bulb of the incorrect rating. A normal flasher relay consists of a bimetallic strip or hot wire to activate the switch contacts. The relay is linked in series with the lamps and the current supplying them is taken through the metal ribbon. The heating effect is controlled by the current through the complete circuit. Thus variations of lamp loading will give variations in current, heating and then flashing frequency.

11 Speedometer and tachometer - bulb replacement

1   The meter illumination bulbs in each of the two meters are bayonet fitting 12 volt 3 watt rating and the holders are held in place by their rubber mountings. They push into the underside of each instrument head.

2   On the S2 model the flashing indicator bulb was fitted to the right side of the tachometer with amber illumination and the blue neutral indicator lamp to the left side of the same meter. The flasher indicator is rated at 12 v 3 watts and the neutral indicator also 12v 3 watts.

3   On S3 models three indicator lamps are mounted between the two indicators. From top to bottom they are:

   Flashers - amber - FL
   High beam - blue - BE
   Neutral - green - NE

4   It is necessary on the UK S2 models to remove the indicators from the mounting plates by unbolting the rubber shock mountings, see Chapter 4, Section 17.5. This gives access to the base of the meters to remove the lamps, their mountings and their associated cables.

12 Horn - operation

1   You are legally required in the UK to have a horn fitted and functioning correctly.

9.1 The stop and tail lamp is replaced by removing the four crosshead screws securing the lens to the assembly

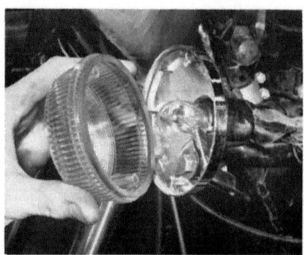

10.2 To replace a lamp remove the plastic lens cover

10.3 Suspect most of all the operating switches on the handlebar assembly

2   The horn is fitted near the crossmember which joins the two front down tubes of the frame below the steering head. It can be adjusted by means of a control on the side.

3   If the horn fails, examine the cables and the horn button connections. The horn button is part of the switch assembly on the left side of the handlebars. Use a meter to see if the switch is functioning and if a voltage reading can be obtained at the horn connections. The horn, turn signal and headlight leads can all be disconnected within the headlamp assembly.

13 Neutral indicator switch - removal and replacement

1   The neutral indicator earthing screw is mounted on the end of the gear selector drum and operates the neutral pilot lamp.

2   If it is not functioning correctly check the applicable lamp in the tachometer and the cable connections with the multimeter before gaining access to the switch itself.

3   The neutral indicator switch is mounted on top of the crankcase inboard from the engine main sprocket. Push down on the terminal to release the cable. Ensure the terminal and cable end is not fouled up with oil or grease.

4   A crosshead screw holds the switch to the crankcase which when removed enables the switch to be withdrawn.

5   Examine the metal earthing strip on the neutral switch and clean it. It should earth on the screw on the gearchange drum when the gears are in neutral.

14 Ignition and lighting switches - general

1   If the ignition switch fails to function a replacement must be obtained as no repair is possible.

2   The switch is located between and below the speedometer and tachometer on the same mounting plate and the two crosshead screws on either side of the switch must be unscrewed for removal of the switch. Be sure to disconnect the ignition switch leads at the connectors first.

3   The old ignition key must be retained as it operates the steering column lock. You will have to keep both keys on your chain.

4   To remove the handlebar switches, first disconnect the battery. The switches are mounted on the left grip assembly and can be split by removing the securing crosshead screws. It should be replaced as repair is not advisable. The assembly contains the horn button, the left/right switch, the headlamp beam dipswitch,

the side light switch.

15 Stop lamp switch - adjustment

1   The stop lamp is operated by a stop lamp switch on the right hand side of the machine just above the brake pedal. It is connected to the pedal by an extension spring, which acts as an operating link. The other end of the switch is connected by cable to the stop lamp bulb. The body of the switch is threaded so that a limited range of adjustment is available, to determine when the lamp will operate.

2   To adjust so that the lamp will operate after the brake pedal has been depressed by about ¾ inch it is necessary to hold the switch body and adjust the nut on that body. Turning clockwise on the nut will make the lamp operate earlier.

3   Our illustration shows the lamp spring about to be reconnected after refitting of the brake pedal.

4   The inbuilt switch in the front brake cable also operates the stop lamp when the front brake is applied. As yet, this facility is not available on UK imported models.

13.3 The neutral indicator switch is mounted on top of the crankcase

15.3 The lamp spring about to be reconnected

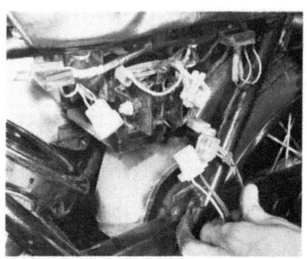

16.1 Where socket connectors are used

16 Wiring - layout and examination

1   The wiring harness is colour-coded and will correspond with the accompanying diagram. Where socket connectors are used they are designed so that reconnection can be made only in the correct position.

2   Visual inspection will show whether any brakes or frayed outer covering are likely to give rise to short circuits. Another source of trouble may be the snap connectors and sockets, where the connector has not been pushed home fully in the outer housing.

3   Intermittent short circuits can often be traced to a chafed wire which passes through or is close to a metal component such as a frame wire. Avoid tight bends in the wire or situations where the wire can become trapped between casings.

17 Fault diagnosis

Symptom	Cause	Remedy
Complete electrical breakdown	Fuse failure Isolated battery	Check wiring and electrical components before fitting a new 15 amp fuse. Check battery connections - do they show signs of corrosion. If so clean thoroughly and apply a thin coat of vaseline.
Dim lights, horn not working	Flat battery	Remove and recharge battery. Are you charging correctly on the machine? A voltmeter (0-20v DC) across the battery terminals with engine running at about 5000 rpm should read 14 - 15 volts. If it does then alternator and rectifier are OK.
Constant fuse failure	Vibration or poor earth connection	Check whether bulb holders are secured correctly. Check earth return or connections to frame.

Ignition Switch Connections

	Batt	Coil	H.L.	Tail	C.L.
OFF					·
ON	●	●	●	●	
CITY LIGHTS	●	●		●	●
PARK				●	·

* Key removable

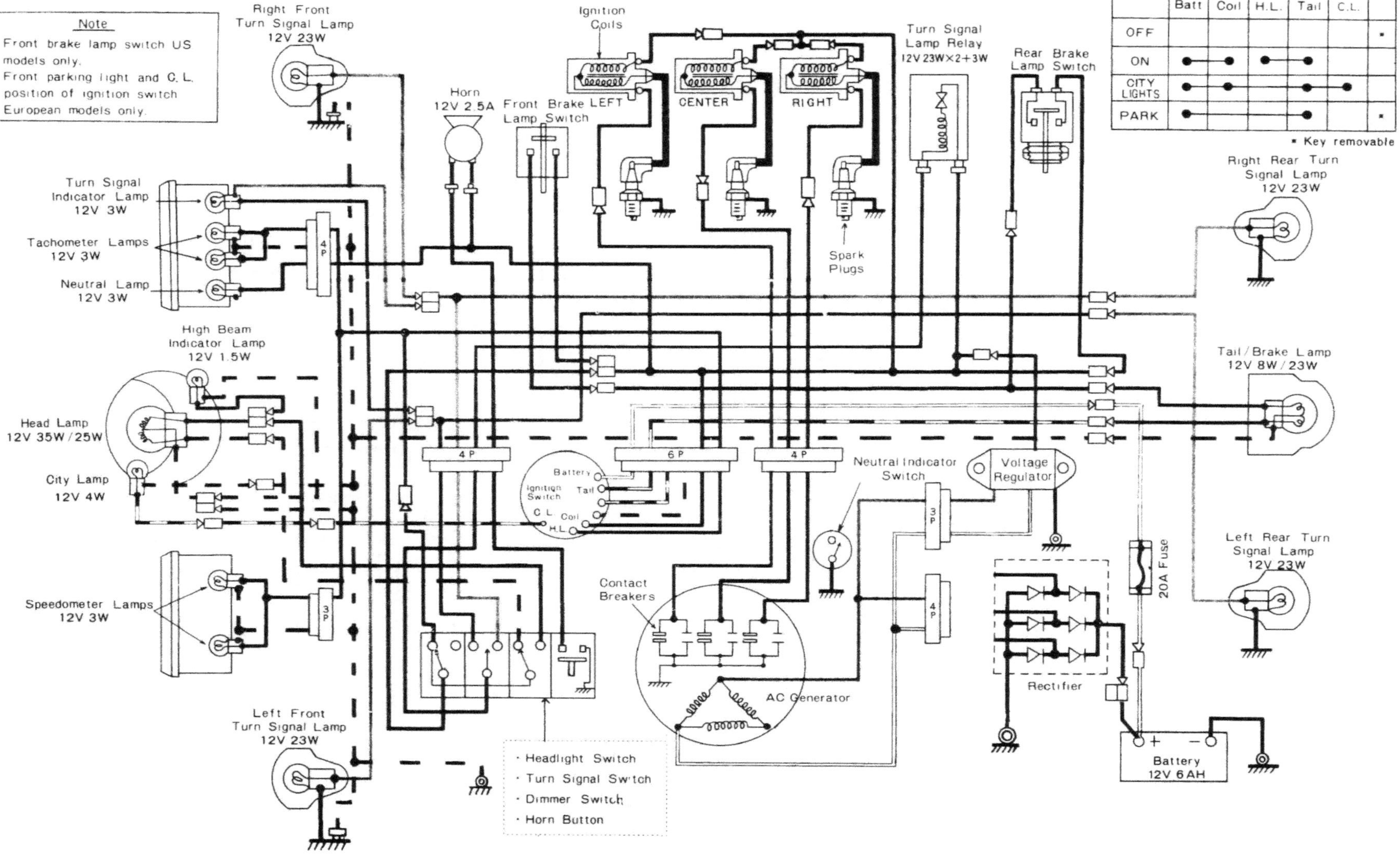

Wiring diagram for S1, S2 (1972 - 73)

Right Rear Turn Signal Lamp 12V 23W

Tail/Brake Lamp 12V 8W/23W

Left Rear Turn Signal Lamp 12V 23W

Battery 12V 6AH

20A Fuse

Rectifier

Rear Brake Lamp Switch

Voltage Regulator

Neutral Indicator Switch

Turn Signal Lamp Relay 12V 23W×2+3W

Spark Plugs

RIGHT

CENTER

AC Generator

Contact Breakers

Ignition Coils

Front Brake LEFT Lamp Switch

· Headlight Switch
· Turn Signal Switch
· Dimmer Switch
· Horn Button

Horn 12V 2.5A

Battery Ignition Switch Tail C.L. Coil H.L.

Right Front Turn Signal Lamp 12V 23W

Left Front Turn Signal Lamp 12V 23W

Turn Signal Indicator Lamp 12V 3W

Tachometer Lamps 12V 3W

Neutral Lamp 12V 3W

High Beam Indicator Lamp 12V 1.5W

Head Lamp 12V 35W/25W

City Lamp 12V 4W

Speedometer Lamps 12V 3W

Note

1. Front brake lamp switch US models only.
2. Front parking light and C.L. position of ignition switch European models only.

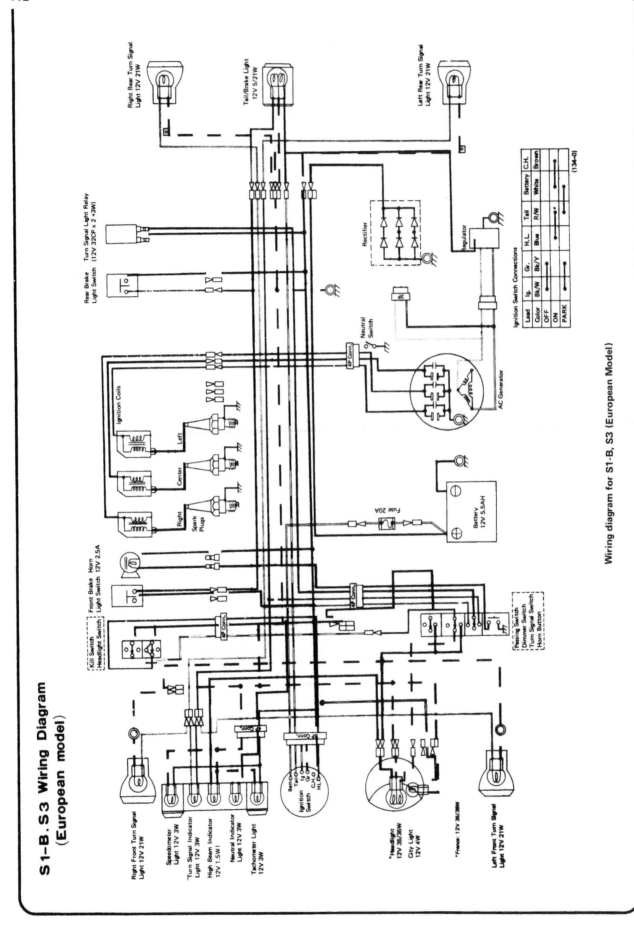

S1-B. S3 Wiring Diagram (European model)

Wiring diagram for S1-B, S3 (European Model)

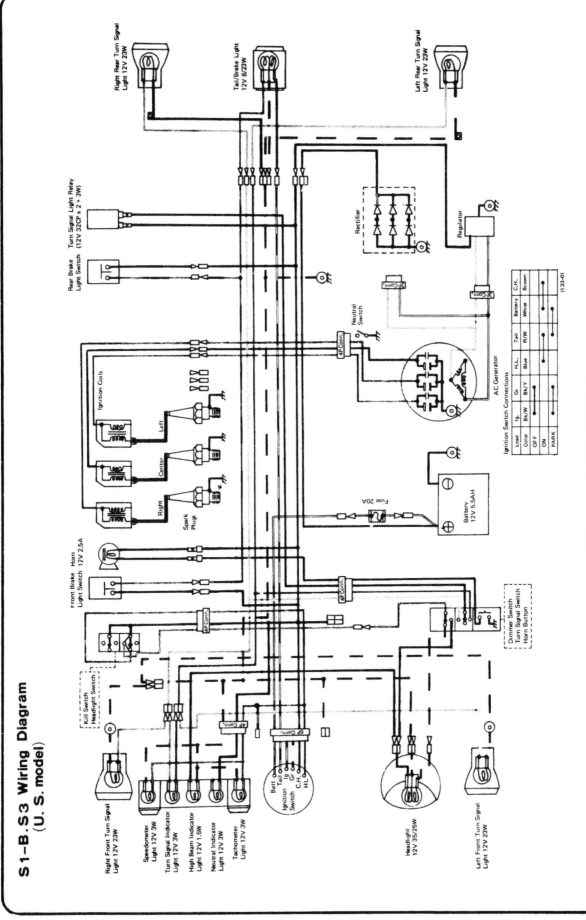

S1-B. S3 Wiring Diagram (U. S. model)

Wiring diagram for S1-B, S3 (USA Model)

Chapter 7: The KH250 & KH400 models

Contents

1 General description

The KH250 Model, which is a successor to the 250 S1 Series, was introduced in 1975 to bring this model into line with the KH400 Model, the updated 400 S3 which was an enlarged and modified successor to the 350 S2. Both the KH250 and the KH400 models are similar in most respects to their predecessors, the main modifications being to the cycle parts and styling.

The design features which differ materially from the earlier models are as follows.

a) Lockable, quick release fuel filler cap.
b) Air filter accessible from the right-hand side of the machine (400).
c) CDI electronic ignition system fitted in place of the earlier triple contact breaker assembly (KH400 model).
d) Design of front fork legs altered internally to incorporate removable fork damping mechanism (KH400 model).
e) Friction type steering damper no longer fitted (KH250 model).
f) Position of steering lock altered (KH250 model).
g) Disc front brake fitted in place of twin leading shoe

drum brake (KH250 model).

Apart from these innovations and certain changes in decorative finish, the information contained in the preceding Chapters can be applied generally to the KH250 and KH400 models. This Chapter covers only the fundamental differences not dealt with in the preceding Chapters.

2 Lockable fuel filler cap

1 The fuel filler cap is of the flip up quick release type, which hinges from the forward edge on a roll pivot pin. The cap is fitted with a cylinder lock, the key of which also fits the steering lock as standard.
2 If the filler cap develops a leak, the gasket may be prised from position and a new one fitted.
3 The lock is not supplied separately to the cap and consequently if it malfunctions, the complete cap assembly must be renewed. The cap is retained in position by the roll pin on which it hinges, which can be tapped from place, using a suitable drift. If a new cap is required, it may be possible to obtain one that the original keys will fit.

3 Air cleaner element (KH400 model only)

1 The air cleaner box is located below the nose of the dualseat. Access to the element may be made after removing the right-hand side cover from the frame. The side cover is a press fit, retained by two projections passing through rubber grommets on the frame.

2 Unscrew the wing bolt in the centre top of the air cleaner box and remove the box cover, which hinges at the forward edge. The filter element, which is of the corrugated paper type, will lift out.

4 CDI ignition: description (KH400 model only)

Unlike the preceding models in the S Series the KH400 is fitted with a CDI (capacitor discharge ignition) system. In common with many other ignition systems, the CDI utilises a rotating magnet/fixed coil magneto to produce AC current for the ignition source power. In this system, however, the mechanical contact breaker unit is replaced by signal coils, which control the point at which the ignition spark occurs. Three signal coils are utilised, spaced 120° apart on the stator plate, each controlling a separate ignition secondary coil and spark plug. As with any other magneto, exciter coils are used to generate the primary (low tension) ignition current.

By replacing a mechanical unit, ie the contact breaker assembly, with a non-mechanical electrical unit, the accuracy of ignition timing is increased and wear is eliminated. Furthermore, once set, the ignition timing remains constant and should not require adjustment.

5 Ignition timing: checking and resetting (KH400 model only)

1 Because of the absence of a contact breaker assembly, the precise point at which ignition occurs cannot be ascertained when the magneto is in a static condition. Ignition timing must be checked with the engine running, using a stroboscope lamp connected in series with the HT side of the ignition system.

2 After removal of the magneto cover, which is reatined by three screws, it will be noted that a scribe line is positioned on the outer perimeter of the flywheel, and a similar index mark is scribed on the stator plate to the left of the upper fixing screw. When the engine is running at 4,000 rpm, the two marks must align in the light from the stroboscope lamp, if ignition timing is correct.

3 Adjustment of the timing may be made by rotating the stator within the limits of the elongated holes which comprise the stator fixing points. Access to the three retaining screws can be made only after removal of the flywheel assembly. This requires the use of Kawasaki service tool No. 57001-116, which consists of a threaded shaft and 'T' handle. If this is not available, an ordinary metric bolt of the required size will be sufficient.

4 During manufacture, after the ignition timing is set, an additional set of index marks are placed on the stator and on the crankcase. These marks are positioned close to the stator upper fixing screw and should be aligned as a method of approximate ignition timing, whenever the stator has been removed and replaced. Ignition timing must then be checked with the engine running, using a stroboscope lamp.

6 CDI unit: location and renewal (KH400 model only)

The CDI unit, which comprises the various transitorised components used in the ignition system, is fitted below the fuel tank. The unit is of the black box type, being permanently sealed and as such must be renewed if any of the components within malfunctions.

7 Front forks: dismantling (KH400 model only)

1 After removal from the machine, the two fork legs may be dismantled and examined. It is advisable to dismantle each fork leg separately, using an identical procedure, to prevent the accidental interchange of components.

2 Place the lower end of the fork leg in a suitable container and remove the drain plug, which is located on the outer face of the lower leg, directly above the spindle recess. Pump the fork leg up and down to aid the expulsion of damping fluid.

3 Loosen and remove the cap bolt from the top of the upper tube (Stanchion) and pull out the spring spacer, spring upper seat and the fork spring. Invert the fork, and using a socket key, remove the socket screw, which is located in a recess above the wheel spindle cut out. This screw retains the damper rod in place.

4 Prise off the dust excluder from around the top of the lower fork leg and slide it off the stanchion. Grasp the lower leg and draw the stanchion from position, complete with the damper rod assembly.

5 Remove the damper piston circlip from the bottom of the stanchion and pull out the piston, complete with the damper rod and valve components. The valve components can be removed from the damper rod for cleaning. Carefully note their relative positions to aid replacement.

8 Front forks: examination and reassembly

1 Examine the outer surfaces of the stanchion for damage in the form of scoring or pitting. This would cause leakage at the oil seal. Check the fit between the stanchion and the fork lower leg. If wear is apparent, these two components must be renewed as unlike the forks fitted to earlier models, separate and renew-able lower leg bushes are not utilised. If slackness occurs between these components, handling problems may develop. Slack forks are also a prime cause of D.O.E. failure.

2 After extended service, the fork springs will take a permanent set (shorten), causing a lower fork level and reducing the spring rate. Measure each spring which should be 10.18 in (158.5 mm) long. If either spring is less than 9.76 in (248 mm) long, the springs must be renewed as a pair.

3 The fork stanchions (upper tube) are unlikely to bend, unless the machine is damaged in an accident. Any significant bend will be detected by eye, but if there is any doubt about straightness, roll the tubes on a flat surface. If the tubes are bent, they must be renewed. Unless specialised repair equipment is used, it is rarely practicable to straighten them to the necessary standard.

4 If there was any evidence of leakage from around the oil seal in the top of either fork lower leg the seal must be renewed. The seal can be prised from position, after removal of the large internal circlip and the backing ring. Removal of the circlip and the oil seal will invariably render the oil seal useless for further service, therefore do not remove it unless a new component is to be fitted.

5 The dust seals must be in good order if they are to fulfil their proper function. Renew any that are split or damaged.

6 Before reassembling the fork legs each component must be cleaned thoroughly in petrol and allowed to dry. It is important that no foreign matter be allowed to adhere to the components during replacement or rapid wear will result.

7 Lightly grease the oil seal and drive it into position with great care, using a suitable drift. Refit the backing ring and circlip. Reassemble the fork legs by reversing the dismantling procedure and then fit them to the machine.

8 The fork legs must be refilled with the correct quantity of damping fluid before the cap bolt is refitted. This is most easily accomplished before the legs are replaced on the machine. Make sure that the drain plugs are replaced and that their sealing washers are in good order.

9   Replenish each fork leg with the recommended amount of
ATF or SAE 10W oil. KH 250 A5 and B1 and KH 400 A3 model,
fill with 212 cc of oil and the KH 400 A4 fill with 208 - 216 cc
of oil.

9   Steering lock: location and replacement

1   The steering lock is positioned centrally in the left-hand side
of the head lug. The steering lock keys also fit the locking petrol
cap as originally supplied.
2   If the lock malfunctions, repair is impracticable and the
complete unit must be renewed. The lock cylinder is retained
in position by the rivet which holds the cover plate. Removal
of the cylinder requires that the rivet be drilled out.
3   Maintenance of the lock is limited to lubrication of the
cylinder using a light universal oil. DO NOT lubricate at the key
hole, only around the periphery of the cylinder.

10 Fuse: location

1   A bank of fuses is located within a small box below the
left-hand side cover. The box contains three 10 amp and two
20 amp fuses, one each of which rate is supplied as a spare.
2   The fuses are incorporated in the electrical system to give
protection from a sudden overload as may occur during a short
circuit. If a fuse blows, it should be renewed, using a spare,
after checking the circuit in question to ascertain the cause
of failure.
3   When a fuse blows and no spare is available, a get-you-home
remedy is to wrap the fuse in silver foil such as that found in a
cigarette packet, replacing it in the holder to reconnect the
circuit. This should only be adopted as an emergency measure,
and a fuse of the correct rating refitted as soon as possible.

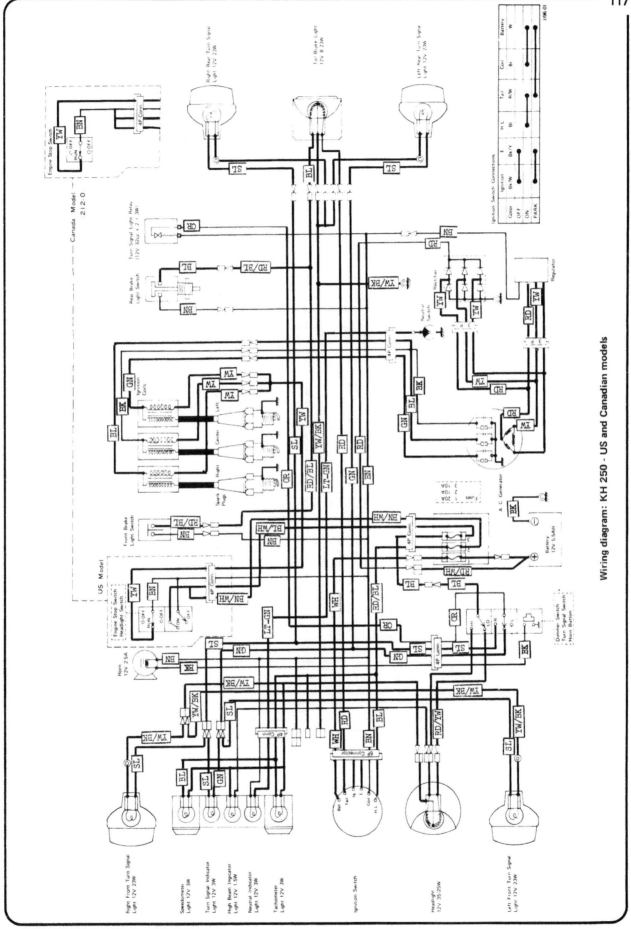

Wiring diagram: KH 250 - US and Canadian models

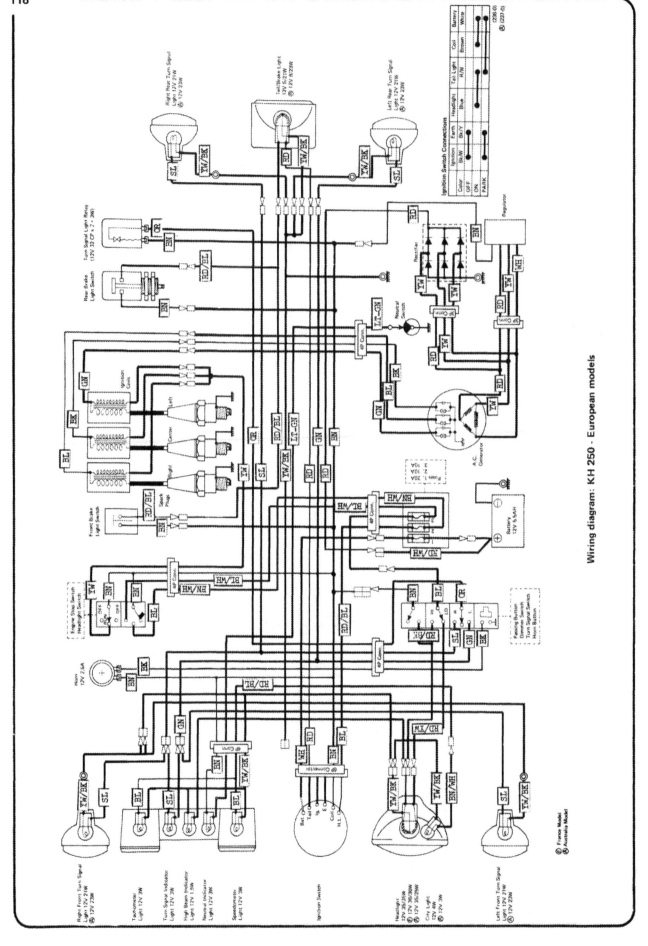

Wiring diagram: KH 250 - European models

Wiring diagram: KH 400 - US and Canadian models

Ignition Switch Connections					
Lead	E	H.L.	Tail	Battery	Coil
Color	BK/Y	Blue	R/W	White	Brown
OFF					
ON					
PARK					

(194:1)

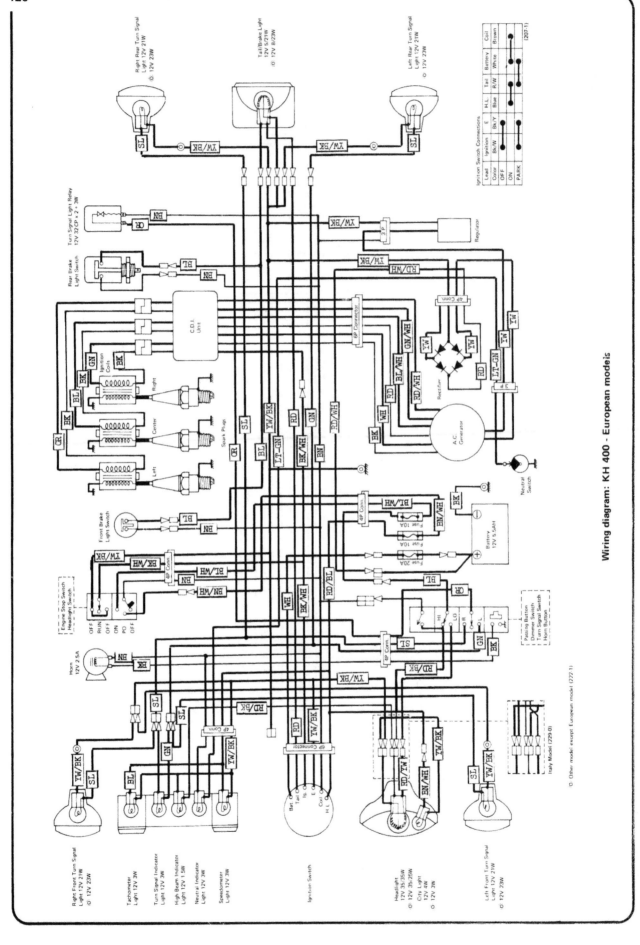

Wiring diagram: KH 400 - European models

Metric conversion tables

Inches	Decimals	Millimetres	Millimetres to Inches		Inches to Millimetres	
			mm	Inches	Inches	mm
1/64	0.015625	0.3969	0.01	0.00039	0.001	0.0254
1/32	0.03125	0.7937	0.02	0.00079	0.002	0.0508
3/64	0.046875	1.1906	0.03	0.00118	0.003	0.0762
1/16	0.0625	1.5875	0.04	0.00157	0.004	0.1016
5/64	0.078125	1.9844	0.05	0.00197	0.005	0.1270
3/32	0.09375	2.3812	0.06	0.00236	0.006	0.1524
7/64	0.109375	2.7781	0.07	0.00276	0.007	0.1778
1/8	0.125	3.1750	0.08	0.00315	0.008	0.2032
9/64	0.140625	3.5719	0.09	0.00354	0.009	0.2286
5/32	0.15625	3.9687	0.1	0.00394	0.01	0.254
11/64	0.171875	4.3656	0.2	0.00787	0.02	0.508
3/16	0.1875	4.7625	0.3	0.01181	0.03	0.762
13/64	0.203125	5.1594	0.4	0.01575	0.04	1.016
7/32	0.21875	5.5562	0.5	0.01969	0.05	1.270
15/64	0.234375	5.9531	0.6	0.02362	0.06	1.524
1/4	0.25	6.3500	0.7	0.02756	0.07	1.778
17/64	0.265625	6.7469	0.8	0.03150	0.08	2.032
9/32	0.28125	7.1437	0.9	0.03543	0.09	2.286
19/64	0.296875	7.5406	1	0.03937	0.1	2.54
5/16	0.3125	7.9375	2	0.07874	0.2	5.08
21/64	0.328125	8.3344	3	0.11811	0.3	7.62
11/32	0.34375	8.7312	4	0.15748	0.4	10.16
23/64	0.359375	9.1281	5	0.19685	0.5	12.70
3/8	0.375	9.5250	6	0.23622	0.6	15.24
25/64	0.390625	9.9219	7	0.27559	0.7	17.78
13/32	0.40625	10.3187	8	0.31496	0.8	20.32
27/64	0.421875	10.7156	9	0.35433	0.9	22.86
7/16	0.4375	11.1125	10	0.39370	1	25.4
29/64	0.453125	11.5094	11	0.43307	2	50.8
15/32	0.46875	11.9062	12	0.47244	3	76.2
31/64	0.484375	12.3031	13	0.51181	4	101.6
1/2	0.5	12.7000	14	0.55118	5	127.0
33/64	0.515625	13.0969	15	0.59055	6	152.4
17/32	0.53125	13.4937	16	0.62992	7	177.8
35/64	0.546875	13.8906	17	0.66929	8	203.2
9/16	0.5625	14.2875	18	0.70866	9	228.6
37/64	0.578125	14.6844	19	0.74803	10	254.0
19/32	0.59375	15.0812	20	0.78740	11	279.4
39/64	0.609375	15.4781	21	0.82677	12	304.8
5/8	0.625	15.8750	22	0.86614	13	330.2
41/64	0.640625	16.2719	23	0.90551	14	355.6
21/32	0.65625	16.6687	24	0.94488	15	381.0
43/64	0.671875	17.0656	25	0.98425	16	406.4
11/16	0.6875	17.4625	26	1.02362	17	431.8
45/64	0.703125	17.8594	27	1.06299	18	457.2
23/32	0.71875	18.2562	28	1.10236	19	482.6
47/64	0.734375	18.6531	29	1.14173	20	508.0
3/4	0.75	19.0500	30	1.18110	21	533.4
49/64	0.765625	19.4469	31	1.22047	22	558.8
25/32	0.78125	19.8437	32	1.25984	23	584.2
51/64	0.796875	20.2406	33	1.29921	24	609.6
13/16	0.8125	20.6375	34	1.33858	25	635.0
53/64	0.828125	21.0344	35	1.37795	26	660.4
27/32	0.84375	21.4312	36	1.41732	27	685.8
55/64	0.859375	21.8281	37	1.4567	28	711.2
7/8	0.875	22.2250	38	1.4961	29	736.6
57/64	0.890625	22.6219	39	1.5354	30	762.0
29/32	0.90625	23.0187	40	1.5748	31	787.4
59/64	0.921875	23.4156	41	1.6142	32	812.8
15/16	0.9375	23.8125	42	1.6535	33	838.2
61/64	0.953125	24.2094	43	1.6929	34	863.6
31/32	0.96875	24.6062	44	1.7323	35	889.0
63/64	0.984375	25.0031	45	1.7717	36	914.4

1   Imperial gallon = 8 Imp pints = 1.16 US gallons = 277.42 cu in = 4.5459 litres

1   US gallon = 4 US quarts = 0.862 Imp gallon = 231 cu in = 3.785 litres

1   Litre : 0.2199 Imp gallon = 0.2642 US gallon = 61.0253 cu in = 1000 cc

Miles to Kilometres		Kilometres to Miles	
1	1.61	1	0.62
2	3.22	2	1.24
3	4.83	3	1.86
4	6.44	4	2.49
5	8.05	5	3.11
6	9.66	6	3.73
7	11.27	7	4.35
8	12.88	8	4.97
9	14.48	9	5.59
10	16.09	10	6.21
20	32.19	20	12.43
30	48.28	30	18.64
40	64.37	40	24.85
50	80.47	50	31.07
60	96.56	60	37.28
70	112.65	70	43.50
80	128.75	80	49.71
90	144.84	90	55.92
100	160.93	100	62.14

lb f ft to Kg f m		Kg f m to lb f ft		lb f/in^2 : Kg f/cm^2		Kg f/cm^2 : lb f/in^2	
1	0.138	1	7.233	1	0.07	1	14.22
2	0.276	2	14.466	2	0.14	2	28.50
3	0.414	3	21.699	3	0.21	3	42.67
4	0.553	4	28.932	4	0.28	4	56.89
5	0.691	5	36.165	5	0.35	5	71.12
6	0.829	6	43.398	6	0.42	6	85.34
7	0.967	7	50.631	7	0.49	7	99.56
8	1.106	8	57.864	8	0.56	8	113.79
9	1.244	9	65.097	9	0.63	9	128.00
10	1.382	10	72.330	10	0.70	10	142.23
20	2.765	20	144.660	20	1.41	20	284.47
30	4.147	30	216.990	30	2.11	30	426.70

English/American terminology

Because this book has been written in England, British English component names, phrases and spellings have been used throughout. American English usage is quite often different and whereas normally no confusion should occur, a list of equivalent terminology is given below.

English	American	English	American
Air filter	Air cleaner	Number plate	License plate
Alignment (headlamp)	Aim	Output or layshaft	Countershaft
Allen screw/key	Socket screw/wrench	Panniers	Side cases
Anticlockwise	Counterclockwise	Paraffin	Kerosene
Bottom/top gear	Low/high gear	Petrol	Gasoline
Bottom/top yoke	Bottom/top triple clamp	Petrol/fuel tank	Gas tank
Bush	Bushing	Pinking	Pinging
Carburettor	Carburetor	Rear suspension unit	Rear shock absorber
Catch	Latch	Rocker cover	Valve cover
Circlip	Snap ring	Selector	Shifter
Clutch drum	Clutch housing	Self-locking pliers	Vise-grips
Dip switch	Dimmer switch	Side or parking lamp	Parking or auxiliary light
Disulphide	Disulfide	Side or prop stand	Kick stand
Dynamo	DC generator	Silencer	Muffler
Earth	Ground	Spanner	Wrench
End float	End play	Split pin	Cotter pin
Engineer's blue	Machinist's dye	Stanchion	Tube
Exhaust pipe	Header	Sulphuric	Sulfuric
Fault diagnosis	Trouble shooting	Sump	Oil pan
Float chamber	Float bowl	Swinging arm	Swingarm
Footrest	Footpeg	Tab washer	Lock washer
Fuel/petrol tap	Petcock	Top box	Trunk
Gaiter	Boot	Torch	Flashlight
Gearbox	Transmission	Two/four stroke	Two/four cycle
Gearchange	Shift	Tyre	Tire
Gudgeon pin	Wrist/piston pin	Valve collar	Valve retainer
Indicator	Turn signal	Valve collets	Valve cotters
Inlet	Intake	Vice	Vise
Input shaft or mainshaft	Mainshaft	Wheel spindle	Axle
Kickstart	Kickstarter	White spirit	Stoddard solvent
Lower leg	Slider	Windscreen	Windshield
Mudguard	Fender		

Index